MW00965046

SOULS, SLAVERY, AND SURVIVAL IN THE MOLENOTECH AGE

AN ALIEN'S VISION

by

Lin Sten

PARAGON HOUSE
St. Paul, Minnesota

First Edition, 1999

Published in the United States by
Paragon House
2700 University Avenue West
St. Paul, MN 55114

Manufactured in the United States of America.

Library of Congress Cataloging-in-Publication Data

Sten, Lin, 1944–

Souls, slavery, and survival in the molenotech age: (an alien's version) / by Lin Sten.—1st ed.

p. cm.

Includes bibliographical references.

ISBN 1-55778-779-4 (cloth)

1. Technology--Social aspects. 2. Nanotechnology. I. Title.

T14.5.S745 1999

303.48'3—dc21 99-23314

CIP

10 9 8 7 6 5 4 3 2 1

For current information about all releases from Paragon House, visit the web site at http://www.paragonhouse.com

Dedication

To my mother for encouraging my ability to communicate, my stepfather for encouraging my interest in science, and Reed College for encouraging my first faltering steps in synthesis.

Permissions Acknowledgements

Acknowledgments and thanks are gratefully given as follows:

to the Albert Einstein Archives, The Jewish National & University Library, The Hebrew University of Jerusalem, Israel for permission to reprint a quote and an excerpt from a letter.

to Pantheon Books, a division of Random House, Inc., for permission to reprint excerpts from *The Face of the Third Reich* by Joachim C. Fest, translated by Michael Bullock. © 1970 by Weidenfeld and Nicolson.

to Wiedenfeld and Nicolson Ltd. for permission to reprint excerpts from *The Face of the Third Reich* by Joaquim C. Fest, translated by Michael Bullock. © 1970 by Weidenfeld and Nicolson.

to Harcourt Brace & Company, for permission to reprint excerpts from *Hitler* by Joachim C. Fest. © 1973 by Verlag Ullstein, English translation by Richard and Clara Winston. © 1974 by Harcourt Brace & Company.

to George Weidenfeld & Nicolson Ltd., for permission to reprint excerpts from *Hitler* by Joaquim C. Fest. © 1973 by Verlag Ullstein, English translation by Richard and Clara Winston. © 1974 by Harcourt Brace & Company.

to the Orion Publishing Group Ltd. for permission to reprint excerpts from *The Face of the Third Reich* and *Hitler,* per the preceding acknowledgement entries.

to Doubleday, a division of Random House, Inc., for permission to reprint excerpts from *Engines of Creation* by Eric K. Drexler. © 1986 by Eric K. Drexler.

to Sterling Lord Literistic, Inc. for permission to reprint excerpts from *Engines of Creation* by Eric K. Drexler. © 1990 by Eric Drexler.

to Harvard University Press for permission to reprint an excerpt from *Commissioner Lin and the Opium War* by Hsi-pao Chang. © 1964 by the President and Fellows of Harvard College.

to Science Service for permission to reprint and excerpt from *Science News,* the weekly newsmagazine of science. © 1993.

to Encyclopaedia Britannica, Inc. for permission to reprint an excerpt from "Hypatia" in *Encyclopaedia Britannica,* 15th edition (1992), 6:200. © 1992.

to Grolier Incorporated for permission to reprint an excerpt from the *Encyclopedia Americana,* 1997 Edition. © 1997 by Grolier Incorporated.

to Macmillan Press Ltd. for permission to reprint an excerpt from *Introducing Psychological Research* by Banyard & Grayson. © 1996 by Philip Banyard & Andrew Grayson.

to New York University Press for permission to reprint an excerpt from *Introducing Psychological Research* by Banyard & Grayson. © 1996 by Philip Banyard & Andrew Grayson.

to Newsweek, Inc. for permission to reprint an excerpt from *Newsweek,* 5 May 1997. © 1997. All rights reserved.

to Lester Maddox for permission to reprint an excerpt from his article in the *World Monitor*, February 1996. © 1996.

to *Technology Review* for permission to reprint excerpts from the February/March 1993 issue, published by the Association of Alumni and Alumnae of MIT. © 1993.

to *MicroTimes* for permission to reprint excerpts from the 26 October 1992 issue. © 1992.

to Curtis Moore for permission to reprint an excerpt from his article in *Outside*, June 1991. © 1991.

to Times-Mirror Magazines, Inc. for permission to reprint an excerpt from *Outdoor Life*, January 1991. © 1991.

to The Institute of Electrical and Electronics Engineers, Inc. for permission to reprint several excerpts from *IEEE Spectrum*, August 1992. © 1992 IEEE

to The Institute of Electrical and Electronics Engineers, Inc. for permission to reprint an excerpt from *Computer*, November 1992. © 1992 IEEE

to U.S. News & World Report for permission to reprint several excerpts from their publication. © April 29, 1996, *U.S. News & World Report.*

to Cahners Publishing Co. for permission to reprint an excerpt from *Variety*, 6-12 January 1997. © 1997.

to Cahners Publishing Co. for permission to reprint an excerpt from *Variety*, 11-17 September 1995. © 1995.

to Cahners Publishing Co. for permission to reprint an excerpt from *Variety*, 4-10 November 1996. © 1996.

to The New York Times Company for permission to reprint an excerpt from *The New York Times Magazine*, December 8, 1996. © 1996 by the New York Times Company.

to Dr. Samuel M. Wilson for permission to reprint an excerpt from his article "Coffee, Tea, or Opium?". © 1993.

to Villard Books, a division of Random House, Inc., for permission to reprint an excerpt from *The Experts Speak* by Cerf & Navasky. © 1984, 1998 by Christopher Cerf and Victor S. Navasky.

to McGraw-Hill Companies, Inc. for permission to reprint an excerpt from *The Western Experience* by Chambers, M.; Grew, R.; Herlihy, D.; Rabb, T.; Woloch, I. © 1995 by McGraw-Hill, Inc.

to McGraw-Hill Companies, Inc. for permission to reprint an excerpt from *Spectacular Computer Crimes* by Buck BloomBecker. © 1990 by Dow Jones-Irwin, Inc.

to the Association for Computing Machinery for permission to reprint an excerpt from *Communications of the ACM*, July 1992. © 1992 ACM, Inc.

to The New York Times Company for permission to reprint excerpts from *The New York Times*, March 6, 1990. © 1990 by the New York Times Company.

ACKNOWLEDGEMENT

The thought and work behind the vision of this book, its writing, and its subsequent publication, are the fruit of many sources, the most significant of which are the Sun, Earth, and Moon, who have generously provided me with sustenance, stimuli, and romance.

More specifically, born out of and shaped by these heavenly bodies are many plants and animals, and Earth's waters, to whom I am deeply indebted for what they have taught me about spirit, science, and sentience.

I am particularly indebted to Homo sapiens for allowing me to learn among them as though I were one of them, and even more so to my mother and step-father, who provided means and encouragement along my path, to my brothers, who provided role models during my adolescent years, and to Reed College for providing a special place of nourishment during my transition to human adulthood.

I also am greatly indebted to Storm for awakening my carnal being; to Diana for a decade of tolerant and supportive companionship; to the University of California at San Diego Physics Department (and my advisor, Professor Norman Kroll) for helping to expand my analytical ability; to my dog Rogue for things he taught me about sight, sound, and smell; to Barney Howard for hiring me into a series of jobs that introduced me to organizational behavior and turned me into a global systems analyst; to Sandra Jensen for her quintessential combination of management skill and analytical capacity; and, to Marcia and her children, with whom my concept of love significantly expanded. Also, I am indebted to an unnamed woman who greatly enhanced my understanding of emotional abuse and human bondage, and whose remnant anger led me to explore my own and that of others. Also important were Nancy Hoshell and Rachel Jacobs for their continuing love and emotional support throughout the writing of this book.

Gloria Leno was particularly important in her reading of an early draft of the manuscript and for providing trustworthy, wise, and insightful remarks thereto.

And most recently, I am thankful to Lillian for her aid and support during publication. At Paragon House, I am thankful to Laureen Enright for her tireless patience in answering my plethora of publication questions and for all the other unseen work she completed in accomplishing the publication of this book, and to Dr. Gordon Anderson for recognizing the uniqueness of this book's vision.

TABLE OF CONTENTS

ACRONYM LIST

AFM	atomic force microscope
AI	artificial intelligence
ATM	asynchronous transfer mode
CPU	central processing unit
DNA	deoxyribonucleic acid
EMR	electromagnetic radiation
IBM	International Business Machines Corporation
Kbps	Kilobits per second (one thousand bits per second)
Mbps	Megabits per second (one million bits per second)
MBP	Munchausen by proxy (Munchausen syndrome by proxy)
mph	miles per hour
NSA	National Security Agency
PKC	public-key cryptography
RNA	ribonucleic acid
RPM	rape, plunder, and murder
RSA	Rivest, Shamir, Adleman
SONET	synchronized optical network
STM	Scanning Tunneling Microscope
U.S.	United States of America
WASP	White Anglo Saxon Protestant
We, we	Depending on the context, this may mean the average reader, average citizen of the U.S., developed nations, or world. (Generally, it means you and I). It is seldom (never) used to mean Homo sapiens as an aggregate.

PREFACE

I was birthed by a human mother. I know not whether the inseminated egg was hers. Neither do I know whether the insemination was natural or artificial, or whether it occurred on Earth. I feel the gestation period was many decades, though I have no memory of my time in the womb. Nor do I remember my birth, though I have papers that imply it occurred on Earth.

Whenever or however I came to be here on Earth, I know that I was raised as a human in a human culture, in a manner that resembles the rearing of most Homo sapiens children in the United States of America. Thus, I learned to enjoy and take for granted all the nutriment, material, social, and attitude benefits that a globally dominant country offers. My salute to the flag was full of pride, and my heart gave all in singing the national anthem each morning at school.

Despite my seeming thorough acculturation into Homo sapiens society, my alien perspective began to reveal itself early, even though at the time I had no idea what I was experiencing since I assumed then that I was a normal human youth.

As a youth I wanted to be accepted and popular like most human children do. Thus, I often attempted to mimic the behaviors around me that seemed to induce acceptance by others. My ambivalence arose with a continuing lack of interest in many of the things that excited other youth: Though I enjoyed the thrill of stepping on the car's accelerator, I had no interest in the camshaft. By the time I completed college I had begun to be aware that much of the discussion taking place around me was boring, and I am now sure that my thoughts were equally boring to my peers. But I still yearned for more acceptance, not knowing that this was a different issue.

For a period of years, during and after my graduate studies, I took an obsessive interest in the meaning of language and how it is used. My focus was on the written and spoken word, without any acknowledged reference to its deliverer or delivery. (In the Aristotelian sense, I focussed on the logos to the attempted exclusion of ethos and pathos.)

During my graduate studies in physics, concurrent screenwriting, and subsequent employment as a mathematics teacher, a systems engineer, and later a global communications systems analyst, I became intimately acquainted with the distinction between logical truth, scientific fact, belief based on comfort, ego based problem solving (in technological and political systems), impartial analysis, individual behavior, organizational behavior, and lynch mob behavior, and truth in its religious, public, and political varieties. Initially I experienced frustration and confusion among so many kinds of truth; I had expected the world to be simpler. My relationships with my family, successive mates, dogs, friends, and other beings also strongly influenced my evolving concept of the universe and how each of us relates to and perceives it. In time my confusion was replaced with compassion.

I gradually learned, as most humans do, how strongly emotional needs drive what we believe to be true. I also became more acquainted with the degree to which peer pressure, and the need for other kinds of approval, guide our beliefs, since most of us very much want to be accepted. That what we believe drives many of our actions, is something we realize at our mother's knee.

Thus, despite my initial scientific orientation and denial of emotion, I gradually came to feel that most of the evolution of society and its behavior is driven by emotion rather than logic. To me it began to seem that strands of logic were often interwoven with other forces motivating aggregate social behavior and the evolution of civilizations, but these strands seemed more like a reinforcement of the final fabric than a substantial part of the design.

I gradually began to sense the danger that lies in the frequent discrepancies between the public's acceptance of ideas and their veracity. Similarly, only slowly within me arose the concept that advances in technology could greatly magnify this danger.

In time too I developed a concept of how deeply in Homo sapiens are buried some of its proclivities. I learned that the brain is formed partly by the identical DNA resident within each of a given person's cells, and partly by experience. It was then that I began to realize that the lessons of evolution, if those are the ones carried by DNA into the primitive recesses of Homo sapiens minds, might not be sufficient to help or allow Homo sapiens to find enough truth for the average person to survive its self-induced radically changing environment. Also, it became clear that whatever the physical mechanisms or divine origins of our existence, we only know that they have been sufficient to allow us survival to the present moment: On a scientific or religious basis, there is nothing about our past that suggests we are biologically, intellectually, emotionally, or spiritually equipped for the future, which will be completely different from anything we have experienced before. It was perhaps acceptance of these ideas that was the biggest step in overcoming my own barriers to consideration of the many possibilities through which I necessarily sifted as the surprising thesis of this book emerged.

This was not an easy book to write. Many times the burden of its harsh bizarre ideas overwhelmed me, tired me, buried me, suffocated me. I often stopped and simply turned away to rest, to write a poem, to enjoy the ocean, to escape. As the book evolved, I began to accept with each departure that I would return, that I would have to return.

In my soul I knew that if promulgation of even the most abominable possible truth could allow this incredible Earth, balanced so delicately between the warmth of the sun and the freezing cold of interstellar space, one sweet extra moment of life, my passion would push me forward.

As the work progressed, I began to see that rather than being pessimistic, this book's suggestions present an opportunity. For Homo sapiens to achieve spiritual evolution, it must be strong enough to absorb unsettling possibilities about its own nature, and to take appropriate precautions, regardless of the risk of failure. Homo sapiens is rapidly approaching a juncture at which this strength will be mortally tested. It is optimism that encourages the feeling that sufficient strength will be found.

Also, during the writing of this book, I was afraid.

My thoughts do not have to make completely sound sense to be significant or dangerous. Indeed, the implied danger exists now, even though many of these ideas could one day prove to be false, and by then the danger could have passed.

It is their possible truth that made these ideas dangerous to me as I wrote. If only some few powerful perceptive people have independently imagined these things, and have already come to believe in their possibility, cabals and dangerous circumstances will arise. The fact that I have imagined these things, and give them credence, is sufficient proof that now or later others will independently do so too.

Consequently, I kept the specific content of this book to myself and several close associates until I put it on the market. Once a degree of public exposure had been achieved, I felt I was somewhat safer.

Now, whether the public responds with rage, mirth, laughter, concern, disinterest, or dedicated change, my relationship to the ideas presented here has altered radically, and is not nearly so private.

At the core of this book is an idea, a suggestion, a possible truth that is disquieting, uncomfortable, dire. Even the temporary consideration of its validity implies a belief that Homo sapiens' negative characteristics could outweigh its positive characteristics. If this is true, our subsequent demise will substantiate it and ultimately stifle any denial; ironically none who remain will recognize the moment of truth.

Some readers may discount this book outright for fear of the images and suggestions.

My challenge to you, the reader, is that you beware of discounting too quickly on the basis of discomfort. Also, I encourage you to not discount the fundamental concept presented herein until you fully comprehend it. After that, I encourage your best judgement.

No slight is intended by my use of "he" in place of "he, she, or it"; convenience is my only excuse, for which I now beg forgiveness if any bias is felt. Further, I intend neither judgement of, nor offense to, any race, religion, organization, gender, creed, corporation, society, or civilization in using behavioral examples. I apologize now for any offense that might later be taken. I have chosen examples of which I am aware, and that I believe are representative of traits of Homo sapiens as a species, or of a majority of its members. If I tend to use a large portion of examples from Christianity, that is due to its familiarity through my Judeo-Christian background.

Lin Sten
Earth
Sunday, July 13, 1997

INTRODUCTION

The story of Adam and Eve in the Garden of Eden illustrates that even the earliest theologians and philosophers knew that Homo sapiens wants continuing enhancements to its survival, and that to achieve these, Homo sapiens will act contrary to spiritual enlightenment, be surreptitious about it, and even when standing next to the omniscient creator will rationalize and attempt to avoid responsibility for it. Regardless of the compassion with which we wish to view ourselves, in the face of rapidly advancing technology, which will soon bring Homo sapiens god-like powers of creation and destruction, our inadequacies may become fatal flaws.

We all wish to see the best of ourselves, and as a society and as a species we look to our corporate, national, global, spiritual, and religious leaders for praise, encouragement, rewards, and guidance. When we fault them for their transgressions, it is always difficult to see that the same faults lie in ourselves with respect to our own analogous behavior toward other innocent beings. For their darker faults, the greatest cover the leaders have is our own reluctance to see that indeed most of us are similar to them, though we are less skilled in their practices.

Through proper guidance, effort, and example, the average human can be taught some measure of morality, and will attempt to live up to this standard, or to appear to live up to it, and to feel guilt in failure. On the other hand, the leaders of the world are the preeminent predator-survivors. They are the ones who develop or encourage moral precepts, profit by them, pay most beautiful lip service to them, avoid them wherever they interfere with power, profit, and pleasure, feel little guilt for their violations, and feel great sorrow, remorse, and embarrassment only when caught in their transgressions. While the leaders may lend cohesion to society in their image of ideals toward which many of us strive, dangerous hallways undermine a society founded on their falsehoods.

In Hitler's Germany during the decade of 1933-1942, a small group of democratically elected leaders was able to gradually take such complete control over another group of people, who were initially free, that by 1941 they were walking them without resistance into gas chambers. That holocaust was not only imposed; it was allowed by those who were compliant and not immediately affected.

If a German citizen, Gentile or Jew, had in 1940 attempted to write and publish a book predicting the holocaust of the Third Reich, what would have happened? If the authorities had discovered the book in development, its author would have silently disappeared along with the book. Would anyone have dared publish such a book? If it had been published, probably the contents would have been ridiculed. Who would have believed that a holocaust was coming? It would have been uncomfortable to believe it; it would have been dangerous to believe it. Which of us reading that book back then would have had sufficient maturity, optimism, and wisdom to face such a threat, instead of conveniently denying it?

On a global scale it is usually difficult to even comprehend what is happening now, much less to see into the future, instead of simply reading about it afterwards. Even if the present were clearly understood, predicting the future is difficult because of unknown variables and the inestimable power of their evolving synergy. Furthermore, acceptance of a prediction about the future encounters additional barriers when discomfort is the companion; happy fairy tales are more easily accepted than disquieting ones.

Regardless, Einstein said, "Great spirits have always found violent opposition from mediocre minds. The latter cannot understand it when a man does not thoughtlessly submit to hereditary prejudices but honestly and courageously uses his intelligence." (Permission granted by the Albert Einstein Archives, The Jewish National & University Library, The Hebrew University of Jerusalem, Israel.) Still, the resistance is bound to be much greater when an idea is unpleasant.

An implied need for action, say to prevent something heinous, lays open the Achilles heel of any proposal that runs contrary to leaving things as they are. The momentum for accepting things as they seem to be is very seductive.

Public policy and peer pressure further dampen acceptance of contrary ideas. An implicit threat discourages too much disagreement or holding the wrong ideas. People do not like change either from their having large material assets, from anticipation of gaining or enhancing their assets, or even from living their simple comfortable lives. It is a clear truth of Homo sapiens that murder is a very acceptable solution for dealing with people who are disturbing the flow of things with uncomfortable ideas and unforeseen difficulties. Whether an innocent messenger slave was bringing bad news to his pharaoh, or Jesus was bringing good news to all, death always lurks nearby for individuals, for societies, for species, and for planets.

Just as the aggregate Homo sapiens population is slaughtering hundreds of species per day as an incidental result of satisfying its wants, so will the ruling class of the near future incidentally slaughter average people with little more than a keystroke. At least a fraction of the ruling class will be fully cognizant, but uncaring, of the incidental deaths they cause; indeed, they will believe what they are doing is necessary for their survival and for the survival of civilization, and they may be right. Paradoxically and unfortunately, our vulnerability will be founded largely on our continuing quest for more and better enhancements to our survival.

For this next holocaust to occur, it is not necessary that all the suggestions or claims of this book be correct. Furthermore, whether or not they are valid, it is not necessary that they be believed. The startling and frightening truth is that the more people reject this book, the more likely its concluding ideas will become true.

In the next ten years, advances in molecular nanotechnology (molenotechnology) and artificial intelligence, civilization's increasing digital orientation, and Homo sapiens' increasing reliance on automated processing and distribution for satisfaction of even basic needs have several revolutionary implications. First, technology is advancing at an increasing rate, and with a sophistication that will make the average human irrelevant to its success. Second, the freedom of the average human will be so limited, and his dependence on processed nutriments so great, as to make him easy to terminate. Third, despite the promise of immortality, the benefits of technological advances will only gradually become available. The evolving ruling class, with god-like powers, will benefit first; indeed, their benefit will come sooner if the rest of us are terminated. Thus, for them the changes will create the opportunity, means, and motivation for a holocaust, and no compunction about it.

It will only take a silent, secretive, determined, miniscule minority to create the next holocaust. Whatever number of us it will absorb, our current ignorance, disinterest, or denial serve to further their clandestine cause.

1

EVOLVING TECHNOLOGICAL CAPABILITY

"Beam us up, Scotty!" to the *Forbidden Planet*

Occasionally it is stated that the technological advances imagined by science fiction writers ultimately come to pass.

This is certainly true of the predictions of Jules Verne. We now have submarines more capable than his *Nautilus,* and we have been to the moon. Mars is next.

As for H.G. Wells, the engineering principles of time machines, at least for time travel in one direction, are now well understood through Einstein's laws of relativity. Actually, time machines exist in nature in the form of black holes. A person only has to go close enough, but not so close as to be ripped apart by the enormous gravitational gradients, and to return to find all one's friends long gone.

In 1939 at the Golden Gate International Exposition on Treasure Island in San Francisco Bay, the average citizen had his first opportunity to be viewed by an electronic camera and to see his own crude live dynamic image displayed on a television screen in real-time. Subsequently, it would not have been a large leap of creativity or faith to imagine that one day the data for the complete atomic description of any given human could be similarly transmitted at light-speed. Thus, by a further extrapolation one is led to imagine a light-speed transport device.

The engineering principles for a possible construction of the transport device shown in the Star Trek television series are quite simple. (See the section *Transport of Mind, Body, and Soul.*) Furthermore, the possibility of constructing such a device depends on technical problems whose resolution is expected within a decade.

These and other mind shattering technologies of awesome power will soon become available.

Whatever structure that can be designed at the molecular or atomic level, and does not violate physical laws of stability, can conceivably be constructed at a moment's notice if the design and materials are provided. Atomic microscopic robot assemblers will be capable of any specific assembly. (For human assembly these assemblers already exist in the form of ribosomes, but these will be surpassed.) Each new design will demand the a priori creation of appropriate assemblers, but that too will soon be possible. Likewise, the disassembly of anything that exists will be as easy and quick.

Anticipation of Homo sapiens coming into possession of these god-like powers of creation and destruction might be accompanied by a sense of jubilation. Control over one's environment has always had a magical allure, and the prevailing attitude is often that good works will result from powerful new tools. On the other hand, good or bad depends on how the tools are used.

With significant impacts looming, the reader may already surmise that, though mind shattering technologies of awesome power will soon exist, serious questions remain as to who will be allowed to use them and who will have control over their use. Other serious questions will arise, and to any of these there may be no easy answers before the technologies and their effects begin to overwhelm us.

Even as we begin to realize that it will not be safe to allow these awesome miraculous powers in the hands of everyone, we may wonder whether we will be safe with these powers in the hands of anyone.

Equally significant to our civilization's possession of these omnipotent powers is the incredible change such a technology will bring to our viewpoint of ourselves as digital beings in a digital world. Data will be furthered in their ascendance as the most meaningful, the most real, things—even more real than flesh and blood beings—and they will become the entities to which most of us will give our greatest credulity and commitment. Furthermore, as individual citizens and as a population, we will increasingly be seen as simply so much statistical data by the ruling class.

The civilization of the Krels in the science fiction movie *Forbidden Planet,* starring Leslie Nielsen (of *Naked Gun*), offered the aforementioned powers and a bit more—the ability to create whatever you imagined just by imagining it. What an exciting fun toy that capability would be! It destroyed them. In fact, we are not so far from that civilization ourselves, if we last that long.

Artificial Intelligence (AI)

In Kipling's *Jungle Book* the jungle boy, Mowgli (Little Frog), translates for his girlfriend what a monkey is saying: "We, the monkey people, are the greatest folk in all the jungle. We know this is true because we always say it is true." Perhaps Kipling was poking fun at humans.

Movie star Tom Selleck related a story about once walking down a street in Hawaii and "feeling full" of himself. Ahead of him were a conjugal pair of tourists. When they saw him, they hurried forward, holding out their camera. He smiled receptively and began to pose for them; they asked him if he would be kind enough to take their picture. He obliged, and in a moment he recognized how full of himself he was, and accomplished a course correction. It is hard to maintain a humble attitude when all activity normally swirls about you.

Some years ago, on the cover of *Life* magazine, there was a picture of a human brain, and inside was the claim, "The [Human] Brain: It is the most highly organized bit of matter in the universe" (Hope). The ignorance manifested in this quoted bit of hubris was then and still is mind boggling, but it is understandable given that most Earth activity swirls around humans.

The human brain is a soggy mess of neurons with dendrites

and axons going every which way to connect with other neurons. Any crystalline object, such as a grain of common table salt, is more organized by far.

Nonetheless, out of a mess of neurons, as forms the brains of many species, comes a great capacity for organization. Anyone who has ever watched and considered for a moment a daddy long-legs spider in motion can guess its minute brain does an incredible amount of processing and organizing. When a flying insect approaches and lands vertically on a wall, its brain is doing significant rapid processing.

Whether or not the human brain is structured in the most efficient manner compared to other known species, or whether it is even the most intelligent, is not yet known. Some scientists have proposed dolphins as the winners; certainly some dolphin species have a greater brain mass than humans. Perhaps more relevant than the size of the brain, which is gigantic in elephants and blue whales, is the ratio of the brain mass to body mass. In this respect, though most dolphin species are comparable to Homo sapiens, crows and vampire bats easily surpass us. Dogs and many other animals have an olfactory lobe that is much more advanced than that of humans. Comparison in the vast remainder of the universe's extraterrestrial realm must await developments.

In a guest editorial for *Computer* in 1992, Michael Conrad also manifested the tendency of humans to be full of themselves when he wrote, "Humans were the first computers." (Conrad) This statement is so blatantly false that it might be excused as harmless huffing or as being too comical for anyone to believe. Nonetheless, its author meant it seriously.

Spiders, insects, birds, bats, dolphins, and many other creatures with very capable brains have been on Earth for longer than Homo sapiens, whether one refers to the Bible or to science.

Additionally, there is the issue of aggregate intelligence, such as is pertinent to the survival of societies, civilizations, and the global aggregate population of any species. Some ant colonies protect, herd, and milk aphids. There is fossil evidence that millions of years before Homo sapiens a species of bacteria gathered uranium to warm its environment.

For millennia Homo sapiens has prided itself on its intelligence, especially in comparison with other species. Of course, in the mixing bowl of survival, if that is the final test for intelligence, considering Homo sapiens' rather brief span here (whether it be 4 thousand or 4 million years) by comparison with other species, such as sharks, dinosaurs, mosquitoes, bats, etc., a long path yet lies ahead in life's most fundamental intelligence test, on which Homo sapiens may prove itself or fail.

If Homo sapiens is eager for early laurels for its intellectual capacity, more solid ground will be found in an incredible and exciting fact: Homo sapiens has already created artificial intelligence (AI) that surpasses Homo sapiens intelligence in many areas. Examples are numerical computations, checkers, and a strategy game called Traveller TCS. The gradual rise of machine intelligence to threaten, equal, and surpass Homo sapiens at chess is well covered in Levy's article in *Newsweek*.

Machine Deep Blue's May 1997 chess defeat of Gary Kasparov, considered by many to be the greatest chess player ever, was the significant termination in a long held human delusion. In the past, ever since the first completely electronic computer (the ENIAC in 1946), many people—experts, laymen, Kasparov, and his mother—have maintained that a machine would never top Homo sapiens at chess. For example, in the mid-1960s, scientist Herbert Dreyfus predicted that no computer would ever beat a 10-year old. In 1979 Douglas Hofstadter, a professor of cognitive science, in his book, indicated continuing doubts that a computer program would ever beat a human chess champion. That most of us once thought this is not surprising; however, our doggedly clinging to this concept, even after the chess playing status of machines began to improve, is an indication of a common attribute of "superior" races that live in denial of mounting contrary evidence.

Equally telling of our conceit is Hofstadter's reassessing remarks made in years after publication of his book:

> There is a constant sense of encroachment by computers....

> What bothers me is the degree to which something incredibly simpler than our brain is starting to be able to do things that we do.... It's taking away from the complexity of what we really are (quoted by Levy).

Probably the disturbing element for Hofstadter and for many of us is that computers are revealing a portion of our cherished complexity to be superfluous.

Even in the face of such a remarkable turning point as occurred in Kasparov's defeat, many humans still cling to what seems to be unique or special about us, claiming that Deep Blue is not alive and that it is not conscious, as though these present truths make us superior and will be forever so.

As we continue to nervously scramble to identify something—consciousness, sentience, etc.—presumably special about ourselves, something that these other intelligent entities are lacking, and about which we can assume a continuing attitude of superiority, we continue to overlook the inevitable truth: these entities are rapidly evolving toward states of being that we have not yet imagined and will never experience.

Regardless of the meaning of such ideas, in the near future artificial intelligence will surpass Homo sapiens intelligence in all regards.

Even though humans have long denied this possibility and still resist its acceptance, it is happening. Even though it is an uncomfortable thought to many people, it really should be no surprise. There is no law of nature or god that says that the human brain has to work better than a brain designed by humans.

Still to some people such an idea seems illogical, paradoxical, or impossible. But consider the following analogy. The human finger measures about one half inch wide, and the visual acuity of the human eye is about five thousandths of an inch. Nonetheless, with such limitations Homo sapiens has developed the ability to measure with an accuracy of nanometers (10^{-9} meters)—roughly the size of atoms—and smaller.

Another basic analogy is this: With a measuring stick that measures only with an accuracy of one inch, humans can make measurements to an accuracy of a hundredth of an inch. This sounds illogical and impossible. Nonetheless, though this is not usually done with rulers, the equivalent of it is done everyday in laboratory measurements throughout the world, incorporating statistical methods that are understood by scientists.

One thing most engineers know: if they get a chance to examine what has already been made, they will be able to make improvements. Creators of AI have worked hard studying natural brains, simulating brains on computers, and improving on nature's examples.

Today, for some of us to continue to see Homo sapiens brains as destined to be forever superior to AI, is like believing Earth was the center of the universe while living in the time of Nicolaus Copernicus (1473-1543), or like believing Earth was flat while living in the time of Christopher Columbus (1451-1506). The progressive thinkers know better, but the doctrine of the multitude lags somewhat behind. Even today, there is no serious harm that can come from continuing these age-old Earth theories, other than being called eccentric. With regard to AI, however, forgiveness for our errors will not be so endearing or enduring.

In any case, to get a meaningful grasp of intelligence, AI, and Homo sapiens' place on the scale, one may examine three critical areas: (1) intelligence within the environment defined by humans, (2) intelligence within the evolving intellectual technological environment, and (3) survival. AI is becoming intertwined with each of these.

Intelligence within the Human Environment

Usually when humans measure intelligence it is done with a test defined by humans. The test problems concern things that are of relevance to human creativity, comfort, and civilization. This is quite natural, but it also creates a bias.

On a freeway, a person with an IQ of 80 usually does quite well navigating home. Likewise, California Highway Patrol officers, chosen for their IQs of about 90, seem to thrive on freeways. Raccoons do very poorly there. On the other hand, if you strip a man naked and confine him to the raccoon's environment in the middle of winter, the man probably will die.

Also, the human slant on intelligence, relating as it mostly does to the "progress" of civilization, contains an implicit, possibly erroneous assumption: that Homo sapiens is on a survival path.

In the big picture, in the long-term view of activity, creativity, and intellectual capacity, we cannot ignore whether the species survives. It does little good to pride ourselves on our intelligence for creating a rifle, and then accidentally blow our brains out with it. But survival intelligence is dealt with at the end of the AI section, and so we can for the moment focus on the issue of all the things Homo sapiens presently takes pride in doing that seem to relate to intelligence.

The sole relevant point here is, simply to repeat, but now with the clarity of what is meant, that AI is on the verge of surpassing individual and aggregate Homo sapiens in all forms of intelligence that we normally consider meaningful in creative scientific and technological advances, our workaday world, and artistic pursuits. At this time, in many of these areas, humans still guide and lead, but AI/human partnerships are increasing, and increasingly humans will be playing the less significant role in them.

Some people may beg the question by pointing out that AI, being the product of aggregate Homo sapiens effort, and a penultimate achievement of human science and civilization, is in fact a part of the aggregate human intelligence, and thus should not be separately compared against us. This is a valid, beautiful, worthy thought, but is not relevant to the point of this section.

The point is stated above, and its relevance is the changing relative roles of AI and Homo sapiens in future developments of civilization.

The survival intelligence of a raccoon in its own environment is greater than human intelligence in that same environment. We have long prided ourselves on our superior intelligence, and presumed computers would never surpass us. In the environment identified by most people as the relevant environment, computers are now increasingly surpassing us.

Intelligence within the Technological Environment

The evolving technological environment is the one on which human survival increasingly depends. Simultaneously with this change, a greater portion of the intellectual capacity of Homo sapiens civilization is accumulating in the technology, and less in humans.

It is self-evident that humans will have relatively less of what intelligence is defined to be. Another way to put this is that humans will make a smaller contribution to the progress of civilization and the evolution of society. This is obviously true insofar as technological advance is concerned; it is probably true too for artistic developments and other creative activities, assuming that they even continue to have a place in civilization.

Until now artistic creativity was mostly human and subjective in its form and success, but artificial creativity will begin to surpass that of humans. For example, already there are screenplay enhancing programs. (See the *More Leisure Time?* section.) Also, the idea of putting a thousand monkeys at a thousand typewriters for a thousand years to churn out literature at the quality of Shakespeare could be figuratively done overnight by computers, and high quality material could be culled by AI the next day.

Of course, what may never be measurable or clear is the extent to which we humans eventually adapt to what machines offer, rather than the machines offering something that is objectively better. Objectivity is an amorphous issue in the world of creative artistic activities.

More importantly though is the fact that humans will be less and less the ones who define the evolving technological environment. Our global planning will become increasingly dependent on AI. Eventually, it will be the technology itself, and particularly AI, that defines what is relevant to progress and to the issue of intelligence.

At the risk of some repetition, this point is worth expansion. Even if Homo sapiens could have retained intellectual superiority in the environment presently perceived by most people to be the relevant one, the relevant environment is changing, and becoming increasingly defined by and in terms of technology, communications, and data processing. Planning, and the capacity to affect this complex environment, will eventually become completely dependent on logical and technological tools (such as computers) within that environment. Since we (average people) will probably not even comprehend this environment, it is specious to assume that we will continue to control it or direct it.

It should not be a shock to imagine that AI will begin to identify the irrelevance of most humans in the near future. Some people will possibly argue that such an idea is ridiculous since humans will program the planning software, etc.; however, in the next ten years AI will more and more take over even that function. Furthermore, the few elite humans who will be in decision making positions will gradually gravitate, whether consciously or subconsciously, toward decisions and plans that reflect their true feelings (not their public lip service) about the average person, which will increasingly be an attitude of disrespect.

Thus, the relevant intelligence and skills for survival in the evolving technological environment will be very different than those of the past. Nonetheless, this environment is the one in which humans increasingly will have to function in order to survive. From the standpoint of daily individual existence, this important fact could have mortal consequences.

At worst, in this environment, most of us are in as much danger as would be a raccoon on a freeway. Or less severely, we will be just as lost as would be an aborigine who might be lured by the miraculous sights and sounds of a vibrant metropolis, and wander into it, while lacking the skills to comprehend or survive in it.

Perhaps more optimistically, most humans will feel supported and excited by the promises of the near-future's materialistically rich technological environment, which we can imagine will supply all our wants and demand less of us.

In any case, we are gradually becoming useless aliens within this new environment, and there is a hidden danger in that, about which more is written later. (See the *More Leisure Time?* section.)

The Survival Test

On the bigger stage of the universe, and the eternity it may offer, the acid test for intelligence of any species may be whether it survives, or how long it survives if it is necessarily limited to a finite existence.

Regardless of humanity's many incredible creations, inventions, and thoughts, any pride in them may only add to our embarrassment if our manner of existing leads to our demise, or if other unforeseen acts of God intervene to cut our excellence short.

For the dodo bird, the intervening factor was humans.

If humans don't survive, the intervening factor may again be humans.

The universe offers many dangers. Certainly pride, and its consequent blindness, is one of them. Greed is another. Additionally, an unseemly large asteroid might hit Earth. In such an event, any being that was not immediately killed might starve to death in the aftermath's sun-blocking global cloud of dust. Alternatively, an alien race could come here and devour us, using their superior intelligence, the needs of their civilization, etcetera, as justification.

As Earth's natural environment changes, two kinds of attributes will enhance a species' chance for survival: (1) Those species with the technology and production capacity to build artificial environments, for which AI is an assumed cornerstone, and (2) those species with short breeding cycles, because short breeding cycles offer the greatest opportunity for a given species to adapt to a changing ecological environment. (See *The Natural Environment* section.) In this latter attribute, raccoons, cockroaches, and mice easily surpass humans. Of course, within a new, inhospitable environment genetic engineering may allow gene mutations that offer enhanced survival capacity for humans.

We do not yet know if AI will increase our chances for survival or hasten our demise. This uncertainty is magnified by the central role AI will play in the development of molenotechnology.

Miraculous Molenotechnology

Fundamental to the advance of AI is molenotechnology. Molenotechnology, automation, and AI will have synergistic, intertwined roles in their mutual advance.

Introduction to Molenotechnology

Molenotechnology (accent on the first "e"), a contraction of molecular nanotechnology, will have a far greater impact on the planet, and on Homo sapiens' civilization and self-concept, than will human population, use of fire, invention of the wheel, the industrial revolution, and nuclear technology combined.

Nobel laureate Richard Feynman (1918-1988), in an address to the American Physical Society in 1959, was the first person to publicly predict the eventual birth and success of molenotechnology (Appenzeller).

In the last ten years Homo sapiens has become able to see, touch, and place individual molecules and atoms. It was only the absence of these capabilities that prevented human industry from surpassing nature's molenotechnology design, construction, and production capacity. Molenotechnology will soon become Homo sapiens' crowning technological achievement.

Even though all living things are molenostructures, they are not the result of Homo sapiens molenotechnology, since Homo sapiens was not involved in these original life designs. However, in the near future, Homo sapiens technology will design and construct things at the molecular and atomic level, just as occurs in nature, and will do so to atomic precision and using molecular machinery and atomic robots, some of whose cognitive capacities will surpass that of humans. This will be molenotechnology.

At the risk of boring the reader with scientific and logical precision, here is a more technical definition: Molenotechnology is the design and construction of entities that are determined and created by the individual specification of each atom (by its type), and by each atom's placement with respect to atomic bonds with its neighboring atoms, whose type, placement, and bonding has been similarly precisely atomically specified and executed.

To quickly understand the impact of this technology, we may consider the following two quotes: Immersed in a carefully controlled bath of nutrient materials, a molenotechnology seed could grow into a rocket engine, just as a hydroponic tomato plant can grow a tomato from a seed.

> What is the engine like?...It is a seamless thing, gemlike. Its empty internal cells, patterned in arrays about a wavelength of light apart, have a side effect: like the pits on a laser disk they diffract light, producing a varied iridescence like that of a fire opal. These empty spaces lighten a structure already made from some of the lightest, strongest materials known. Compared to a modern metal engine, this advanced engine has over 90 percent less mass (Drexler, 62).

> An invention made on Monday could, by the following Friday, be in mass production, with billions of copies fabricated.... Our economic and societal structures have evolved around assumptions that will no longer be valid once technology reaches this milestone (Walker, 286).

Nature's Validation

Molenotechnology has already been validated by nature herself. Whether the source was accident, cosmic creation, spontaneous generation, evolutionary force, divine will, or alien action, Earthlife is pregnant with proof of the feasibility of molenotechnology.

The distinguishing feature of molenotechnology from what has been occurring for millions of years (or thousands, depending on one's biblical leanings) is only that Homo sapiens is on the verge of superseding these previous forces in the invention, design, fabrication, full-scale production, and replication of molenotechnology entities. Furthermore, by definition they will no longer be necessarily chemically organic in spite of how alive many of them will be. (Nonetheless, human tenure in these activities will be brief, as we too will soon be superceded.)

In nature, each living entity is composed of cells, and the nucleus of each cell of a specific being contains an identical DNA (deoxyribonucleic acid) molecule. The DNA molecule contains the complete design (or blueprint), the entire atomic and molecular specification, for that whole life entity. (See the *Immortality* section for the distinction in the development of a human brain, for example, between what is due to the DNA and what is due to experience.)

Enzymes, hormones, antibodies, and structural components of cells are all protein molecules. Proteins are the fundamental building blocks in cells of all living things. As many as 50,000 kinds of proteins may be used in a single animal cell.

All proteins are constructed by molecular machinery. Within the cytoplasm (outside the nucleus) of each cell, polypeptides (molecular chains of amino acids) are synthesized by atomic robots to atomic precision, using designs encoded in molecules at the atomic level, and whose final product is a specific protein.

Ribosome is the biological name of the atomic robots that exist in the cells of all living things to manufacture proteins. The design for each such protein is contained in a protein specific RNA (ribonucleic acid) molecule, which is duplicated from a portion of the DNA molecule within the nucleus. A ribosome reads a messenger RNA molecule, wherein a specific protein design is encoded, to synthesize a protein from amino acid molecules, one molecule at a time. An amino acid specific transfer RNA molecule delivers each amino acid molecule to the synthesis site, where the ribosome is bound to the messenger RNA. The ribosome then bonds the indicated amino acid to the growing polypeptide, and moves to the next unit of the messenger RNA to read it and repeat the process with the next required amino acid molecule. (Actually in the bonding process each amino acid is stripped of a water molecule, leaving an amino acid residue. These residues constitute the polypeptide.) The ribosome successively reads each unit of the messenger RNA, and bonds the respective amino acid to the growing polypeptide, whose completion is the designated protein (specified in the messenger RNA). (See more information about this in [Phillips] and in [*Aca-*

demic American Encyclopedia, Vol 15, p. 575.])

An example protein is the enzyme molecule lysozyme. Its structure was understood prior to 1966 and was disclosed in an article by David C. Phillips in *Scientific American* in November 1966 (Phillips). Lysozyme contains 1,950 atoms in a structure that basically consists of a single polypeptide of 129 amino acid molecules, of 20 different kinds.

Thus, a redwood tree, a human, and a worm are all examples of molenostructures, in so far as they are all built of cells, and proteins are the fundamental building blocks of cells of all living things, and ribosomes exist in all living things to manufacture them according to the atomic design blueprint given in the DNA. Likewise, a single-cell bacterium and the ribosome are molenostructures.

Current Status

Humans already guide and support the construction of molenostructures through their own procreation and industry. Nonetheless, the methods are still those of bulk technology, which does not design and control manufacturing on an atom by atom basis. Houses, cars, arrowheads, sculptures, photolithography, and etching, are examples of bulk products created by bulk manufacturing technology.

Halobacterium halobium grows naturally in the salt marshes of San Francisco Bay. This microorganism is excitingly unique: when the oxygen content of the marsh becomes too low for adequate respiration, this bacterium switches to photosynthesis for its energy—partially switching from being an animal to being a plant. The protein bacteriorhodopsin is the fundamental component of this organism's photosynthetic activity, and there is a chromophore molecule bound to it. Each chromophore molecule is identical and consists of less than 100 atoms. This chromophore molecule is another example of a molenotechnology structure that is used by humans. Each chromophore molecule can theoretically store one bit of data, and has a very high switching speed—500 femtoseconds (femto means 10^{-15} or one quadrillionth).

Though Homo sapiens did not design this organic molecule, for data storage purposes we encourage its manufacture through bulk methods in laboratories (Birge; Freedman).

There are many other examples of the current status of molenotechnology. Several products and tools are given here.

> In 1988, a group at DuPont led by William DeGrado designed a new protein, called alpha-4, from scratch, and manufactured it in their laboratory (Walker, 128).

The Nobel prize in chemistry was given in 1987 for the design and synthesis of a molecule, albeit bulk technology was used in the synthesis. For this molenotechnology feat, several people shared the prize: Donald Cram of UCLA, Jean-Marie Lehn of Universite Louis Pasteur, and Charles Pederson of DuPont.

The Nobel prize in physics was given in 1986 for the invention of the Scanning Tunneling Microscope (STM). The recipients, Gerd Binnig and Heinrich Rohrer of the IBM Zurich Research Laboratory, invented it in 1981. The STM is a sophisticated combination of hardware, software, and a tungsten needle whose point is a single atom. Though its use is restricted to low temperatures, it can be used to make three dimensional images of atoms, to pick up and move single atoms, to stick single atoms to a substrate, and to break molecular bonds. Since STM imaging depends upon electric current in its needle tip, nonconducting biological molecules must first be coated with a conductor or pinned to a conducting surface (Kinoshita).

As a demonstration of the STM's technical power, IBM (San Jose) used one to individually place thirty-five xenon atoms on a nickel crystal substrate. The resulting construction was a five nanometer high spelling of "IBM," of which the STM also composed the photographic image. (A nanometer is about twice the diameter of a xenon atom.) The picture was shown in *MicroTimes*, 26 October 1992, on p. 123 (Walker).

> John Foster's group (at IBM) has observed and modified individual molecules using the technology of the scanning tunneling microscope (Drexler, 240).

As other examples of the power of the STM, it has

> picked up individual atoms of copper and silver and piled them in four tiny heaps to form the world's smallest battery.... Bent molecules of DNA in a sturdy cube.... Built a tiny switch that flicks on and off with the passage of a single electron (Chui, *San Jose Mercury News*, 6 December 1992).

The Atomic Force Microscope (AFM) can also interact and make images at the molecular and atomic level. It can put reactive molecules together so they can combine. Though it cannot grip a molecule, it can probe the molecule to determine its shape, and make an image of it. Its effective use has been described by Drexler thus: "like a robot arm, able to position things in three degrees of freedom...it's like a robot arm that ends in a rough rock, rather than a selective gripper" (Eisenhart, 179). It has the additional advantage of being applicable to nonconductive matter, whether living cells or molecules.

In one experiment an AFM wielded a catalytic tip to reshape a molecular landscape by moving one molecule at a time (Travis).

The AFM will evolve into a first-generation molecular manipulator. The end of the AFM probe will have a particular protein molecule attached. This protein molecule will allow the molecular manipulator to grip a particular molecule in a particular way, and thus to position molecular building blocks.

Such atomic and molecular manipulation and imaging, as have already occurred, is the precursor technology to that which will be essential to the design and creation of assemblers, which will be discussed below in the *Molenorobots* section.

For several years throughout the world researchers have been designing molenotechnology systems. Likewise, according to Drexler, computer-aided design of such systems has already occurred. The article, "Computer simulates a nanofactory" (Chui, *San Jose Mercury News*, 8 December 1992), shows a picture of a molenotechnology gear system of eleven moving parts and 3,557 precisely placed atoms. This gear system and its manufacture have already been validated by computer analysis and simulation: the gear system works and its manufacture is possible.

> The ability to model and simulate complex molecular systems has been growing rapidly in recent years, driven both by advances in raw computing power, but also by the development of better simulation techniques that now permit modeling of proteins composed of thousands of atoms (Walker, 128).

Evolving Developments

It will be possible to "put [all] the [current] computing power of the world on a single chip" (Hapgood paraphrasing Drexler). Data storage technology will also advance:

> For example, with the information represented by a code with elements composed of clusters of a few atoms, every book, magazine, pamphlet, and newspaper ever printed, complete with graphics, could be stored in a volume the size of a credit card (Hapgood).

> A micron-wide mechanical computer...will fit in 1/1000 of the volume of a typical (living) cell, yet will hold more information than does the cell's DNA (Drexler, 105).

> Even with a billion bytes of storage, a nanomechanical computer could fit in a box a micron wide, about the size of a bacterium (Drexler, 19).

To Eisenhart, Drexler describes computers with "billion-cycle-per-second clock rates...smaller than bacterium...consume so little power that a billion times the capacity of a modern supercomputer could be put in a desktop box and cooled with a fan" (Eisenhart, 134). It is an incidental tribute to Charles Babbage (1792-1871), whose mechanical computer could not be constructed in his era due to technological limitations (but which was constructed recently) that this imagined miraculous molenotechnology computer will also be mechanical.

We may wonder how a mechanical computer can run faster than presently existing electronic computers. The reason is simple: its parts will be atoms and molecules. To understand the significance of size with regard to speed, compare the maximum up and down movement of a human arm, say, once or twice per second, with that of a mosquito wing, at 200-500 times per second (*Time*, 10 August 1992). The smaller the faster.

Artificial intelligence (AI) will be integrated with advanced computers. This symbiosis will be used in the design process, which will use computer-aided molecular design, and in the dynamic simulation of molecular structures, which is necessary for the advance of molenotechnology. In a bootstrap process, AI will be enhanced, and it will subsequently enhance the development of software for the design of molecular systems, for the mapping of their construction, for their simulation, and for predicting their stability and capabilities.

Drexler claims that a cubic centimeter brainlike electronic computer made of molenotechnology components could run ten million times as fast as the human brain (Drexler, 79). Such machines will create designs at heretofore unimagined rates. And in days, rather than in years (as was the case up until now), those designs that are of molecular components could be simulated, turned into prototypes, and tested.

Before rapid prototype building and testing can occur, molenotechnology assemblers must be constructed. These assemblers will be microscopic molenotechnology robots (molenorobots).

Molenorobots

Molenorobots (molenotechnology robots) are robots whose every working part, whose every molecule, whose every atom, is specified in its placement, orientation, bonding, and function. Though such robots will exist in any size, their greatest initial impact will be through those of microscopic size. Nonetheless, many of these molenorobots will have an intelligence far surpassing the most brilliant human.

The development of molenorobots depends upon the creation of several other molenomachines—assemblers, disassemblers, and their associated molenocomputers.

Assemblers are atomic and molecular machines that can read molecular design code, and construct molecular entities by the atomic precision placement of molecules and the forming of specified atomic bonds. Nature has already shown that assemblers can be designed, constructed, and function. Ribosomes, reading RNA and working in conjunction with enzymes, are a good example.

Several quotes from Drexler's book are included below to provide context, and amplify the reader's understanding of the full potential of the creation of assemblers.

> Molecular assemblers will bring a revolution without parallel since the development of ribosomes, the primitive assemblers in the cell (Drexler, 21).

> Able to tolerate acid or vacuum, freezing or baking, depending on design, enzyme-like second-generation machines will be able to use as "tools" almost any of the reactive molecules used by chemists—but they will wield them with the precision of programmed machines. They will be able to bond atoms together in virtually any stable pattern, adding a few at a time to the surface of a workpiece until a complex structure is complete (Drexler, 14).

> Because assemblers will let us place atoms in almost any reasonable arrangement, they will let us build almost anything that the laws of nature allow to exist. In particular, they will let us build almost anything we can design—including more assemblers (Drexler, 14).

> Eventually, assemblers will allow engineers to make whatever can be designed, sidestepping the traditional problems of materials and fabrication (Drexler, 49).

Unlike their natural relative, the ribosome, assemblers will not have to be of organic material, and their products can be organic or inorganic.

A disassembler is the reverse of an assembler. It disassembles molecular entities an atom or molecule at a time, and records their molecular structure in molecular code similar to DNA. Such molecular code can be passed to, and used by, an assembler.

A molenorobot will be a combination of assembler, disassembler, and computer, or several of these components. Each component can be of nanometer size, or much larger. The smallest of such entities will consist of several thousand atoms. They will be capable of performing almost any task that can be imagined. (The implications for medical technology are profound, and these will be discussed in the *Immortality* section.)

Among the most startling capabilities that will soon exist for some molenorobots will be the capacity for replication. They will be able to build duplicates of themselves, regardless of whether such reproduction is sexual or purely mechanical. Such robots can be called replicators.

Eventually some replicators will be self-sustaining. They will not depend upon humans. They will be able to replicate themselves and other things, and survive, whether or not Homo sapiens survives.

Furthermore, they will be able to mutate in ways that they will be capable of deciding for themselves. Though Homo sapiens will create and initially use these replicators, these replicators will be more intelligent than any human, and will soon acquire the capacity to determine their own destiny.

They may decide that Homo sapiens is of no use to their own future. Are we too blinded by our own glory to realize that such replicators may determine that Homo sapiens is a detriment to their future? Might they discard us as we presently do other Earth life?

These replicator entities will be more fit, less vulnerable, and more adaptable, than any known present or past living species.

The laws of natural selection will apply. The strongest will dominate.

[In writing the *Miraculous Molenotechnology* section, the author gratefully acknowledges the use of a 17 December 1993 unpublished paper, "The Impact Of Molecular Nanotechnology On Key Technology Areas," now owned by Lockheed Martin Corporation and which he wrote while he was an employee of Unisys Corporation.]

Immortality

The technology for biological immortality is at hand.

One of the incredible scientific consequences of molenotechnology is that immortality will be possible, as will rejuvenation. Indeed, anyone will be able to attain any age they wish, or even to attain the form of another animal without sacrificing their human brain or basic psychological being and personality. But this is not a biological or genetic engineering revolution, it is more grandly the capability to manipulate atoms, any atoms, and to have themselves build any desired form, or to identify and repair any damaged or diseased form, with atomic precision—molenotechnology.

Life is the most fundamental treasure of every being. Like most of our gifts, for most of our lives we take it for granted as surely as the air we breath. We certainly spend most of the days of our lives following the course set by our primitive motivation to survive (and to enhance our survival), but often we forget to give thanks that we are alive. This gift of life is so magical, and of such great duration compared to a day, that in human youth there is the implicit assumption of immortality: nothing can hurt us and we will live forever.

Nonetheless, at some point in most normal human lives (if uninterrupted by a fatal injury), each of us begins to have a concept of how many years we have left before we die. For many people this change of attitude begins to emerge during middle age, or it may be spurred by the death of a parent, which serves as a strong mortality awareness bell. Depending on our religious background, the concept of our spiritual immortality may then begin to fill the void of our vanishing youthful vanity; we may begin to give up our belief in physical immortality.

The biblical promise of immortality has an allure that is hard to ignore, especially since it has more than spiritual implications. Most people who imagine a hereafter do not confine their fantasy to shimmering orbs of angelic light; they think of themselves, as they do of God, in a human form. And of course Christians were not the first to espouse such anthropomorphic immortality.

Ancient Egyptians believed so firmly in immortality that they mummified their dead. Additionally, at least for the wealthy, contracts were made to supply the tomb with a continuing flow of worldly sustenance. Furthermore, figurines of concubines, whose (planned) purpose is obvious, were often included in the tombs of men. Too, inscriptions on tombs begged the passing traveler to give whatever he could so that the life of the entombed could be eased later.

Tutankhamen's Limited Immortality

Part of the ancient Egyptian promise can be kept. The technology exists to biologically resurrect Tutankhamen, at least to a limited extent. Today this can be done through the process of cloning, which is not related to molenotechnology.

The movie *Jurassic Park* explains that a single molecule of DNA contains all the necessary design information for the identical replication of any being. For a given being, identical DNA exists in every one of its cells, as was stated in an earlier section. Thus, a single cell, whether it be a sluffed skin cell, a cell of hair, or of intestine, etc., would contain the necessary DNA. Of course, when a mummified corpse (or dinosaur) is in question, one must find a strand of DNA that still exists in its entirety, or one must be able to construct it from various pieces (as in *Jurassic Park*). If sufficient DNA exists then cloning can replicate the (nearly complete) biological living person, but this does not include the person's memories or personality.

Cloning is a process in which a single DNA molecule is used, presumably within some suitable egglike environment, to naturally grow a replica of the DNA donor. Such growth takes place at nature's rate, and takes the growing organism through all the stages of biological growth. For a human these stages would include fertilized egg, embryo, fetus, baby, adolescence, etc.

In this manner, the physical being of Tutankhamen could again walk this earth (assuming the availability of a single intact DNA molecule), but the person will not; neither his personality nor memories can be cloned or reconstructed. Indeed, in Tutankhamen's time, the brain was removed from the corpse with an iron hook inserted through the nose. So this new being would know nothing of ancient Egypt; he would merely be a biological replica with a brain developed by his experiences in this present life. Thus, the ancient Egyptians were right in their belief in immortality, but their memories and personalities have been severally left behind in the dissolving fluids of their canopic jars, into which their brains were plopped during mummification.

It is likely that many members of the ruling class already are in possession of clones, which have their identical DNA. This is technologically and medically possible. These clones will supply them with young body parts to which their bodies will not have any immune response. Ethical issues presently keep this practice away from public view. The people who are doing this know in their souls that the volatile ethical issues that public view would raise are only superficial; they know that the core of the average citizen's response would derive from jealousy and envy. These people also know that once the possibility of clones in storage is available to everyone then the ethical issue will dissipate.

The central issue is simply the present and growing disparity between the ruling class and the rest of us. It is the same as the disparity between the pharaoh and the average citizen slave: They knew what to do to offer immortality, mummify, etc., but they could not afford to send everyone forward. For us the ruling class knows it is best to keep this particular disparity, between life and death, out of sight for now.

In the era of molenotechnology, DNA based replication of the biological being will also be possible, but it will happen very differently. A DNA molecule will still provide the design, but the construction will be done by assemblers that are of human design, instead of being done by nature's ribosomes. Construction will not necessarily have to proceed through any of nature's growth phases. More efficient construction paths will be found to reach a desired end product. Construction will run quickly, eventually in seconds, to the specified completion age.

Nonetheless, through the methods of molenotechnology, and using only a single intact DNA molecule, a resurrected Tutankhamen would fair no better than his cloned twin in his memories—they are gone.

Rejuvenation and Healing

It takes little additional imagination to realize that in the era of molenotechnology, medicine and the health sciences will be very different from now. The difference will be greater than any comparison could suggest.

Microscopic molenotechnology robots (molenorobots) will aid in prevention of death. Not only will disease be cured, but its damage will be repaired; this will be done in precise, molecule by molecule operations carried out by surgical molenorobots. These molenorobots will not wield scalpels; they will remove and insert specific single atoms or molecules at particular atomic bonding sites.

Take cancer as an example. In scalpel or laser surgery a great deal of healthy tissue must be cut, scarred, or even removed. In chemotherapy, the whole body is poisoned; its use is based on the cancer usually dying faster than the healthy tissue. In molenotechnology cancer treatment the patient will drink some fruit juice containing perhaps millions of sophisticated molenorobots. Through the stomach wall these molenorobots will be absorbed into the blood stream, which will carry them to the diseased region. Their sole function will be to identify and disassemble the cancer cell-by-cell, molecule-by-molecule, atom-by-atom, and to do any necessary repair work to restore the surrounding tissue to good health.

Any disease or injury could be similarly eliminated or repaired, respectively, as long as the robots are given the correct instructions, most of which could be obtained from immediately accessible in situ DNA and the rest of which would be based on the particular person's history.

The technology of rejuvenation will be very similar to that of cure or repair work for any other disease . The critical technology is access through DNA to the precise information of what needs to be done at the molecular level and the existence of molenorobots to do the work.

In fact, rejuvenation to any desired degree will be possible, so that people can choose to stop or reverse their aging.

Likewise, a person could as easily choose to have an older appearance. This choice may be more appealing to us once the stigma of death is removed from age; however, it is possible that most of us would want to choose an age of maximum health and vigor.

In any case, through rejuvenation, cure of disease, and repair of injuries, biological immortality will be possible. (See the *Danger* section for when this will be possible and other relevant details.)

Resurrection

With molenotechnology we will literally be able to raise the dead.

The methods of rejuvenation discussed in the previous section could be used to resurrect someone who had only recently died. The brevity of time subsequent to death would be determined by the necessity of the deceased person's brain still being intact. The brain must retain enough integrity so that the neural structures that are unique to that person's personality and memories would not be lost before regeneration and repair of damaged tissues.

In the case of violent or accidental death, resurrection only depends on getting the dead body to a medical center where the damage can be repaired and the body functions restarted. As long as the specific form of the brain has not been destroyed, the complete person can be restored and reanimated, and they will know who they are—their memory and personality will be intact.

The organic integrity of the brain is important since, as suggested in an earlier section, memory and personality do not reside in DNA. There is strong evidence that the particular form of each human mind is not determined specifically by its DNA; rather, experience and environment play a role in the neuron-by-neuron formation of the brain. Also, any other peculiarities of one's body, such as scar tissue, moles, warts, etc., though possibly undesirable, are not dictated by DNA.

Underneath our skulls, the surface of the brain and its gross form may in its appearance be largely determined by our DNA just as our faces are. On the other hand, at the microscopic level, among the neurons the structure is not so simply determined. Which neurons are joined to which is very much determined by our experiences, and is intimately connected with our memories and personalities.

Even in the womb, fetuses begin experiencing things. They hear music and voices, feel the pressure of something resting on their mother's lap, and possibly even experience some of their mother's emotional state. From these earliest experiences, the development of our brain begins to blend with the course set for it by the DNA residing in each cell of our body.

After birth, our brains continue to develop very much according to our experiences.

Barring any serious accidents, diseases, or cosmetic surgery, our bodies are formed mostly according to our DNA, and somewhat by our experiences. For the development of our brains, our experiences have a greater influence.

Thus, after a person's death, if the body and brain are still whole, then through molenorobots all the relevant information can be obtained from within the body and brain, and complete reconstruction is possible.

In the initial fruition of this capability, the reconstruction of a person might be quite slow, but as the technology advances reconstruction could take place in several seconds. (The same would be true for rejuvenation.)

What about resurrection long after death, after the deceased person's brain has decayed?

As discussed above, to recreate a whole person—body and mind—takes two kinds of information: the blueprint contained in a DNA molecule and a complete neuron map of the mind. If all this unique information is available, resurrection is possible.

How would a living person get their complete body/brain map on file? Millions of molenorobots, each consisting of a disassembler, reader, assembler, and computer would infiltrate the person. In tissues and cells throughout the body and brain, these molenorobots would disassemble and re-assemble at the cellular and molecular level to read, map, and store all the information.

To get some idea of how much space might be needed to store an individual map, consider the following. Even though the cells of any living being contain over 50,000 distinct proteins, a DNA molecule stores all of this information and most of the remaining relevant information very compactly: it is microscopic. The map of the human brain probably will not have to be at the molecular level, except for the molecular map of a single neuron-to-neuron connection; other than that, probably a neuron level map will suffice, showing all the neuron-to-neuron connections, and this could be stored in far less than one thousandth the volume of the human brain. So, one may easily surmise that with future molenotechnology, only microscopic space would be needed to store the data for a complete specific human.

Of course, in this imagined near future, a person of the constant guaranteed age of 24 might still step in front of a speeding semi-trailer. Or a gun shot will still kill. Subsequent to having been mapped and stored, if the person were killed in a car accident, for example, even cremated, he could be regenerated and reanimated from this stored information, and he will know who he is—his memory will be intact.

The technology to store the complete information of a human being also allows the fascinating medical development of operating on a cancer (or performing rejuvenation) by operating digitally on the person's data file, and regenerating the cured person from the corrected file. The previous body would legally agree to termination in consideration for this digital renewal service.

From a philosophical standpoint, perhaps the most significant issue here is that the complete data for a person's regeneration, of mind and body, could be stored on disk, and that person could be regenerated at any time. This certainly will bring a new understanding of the meaning of life and death, and raise questions about the residence of the soul.

Transport of Mind, Body, and Soul

With a complete mind map and DNA map on file for a given person, any number of living replicas could be produced at moment's notice. Those replicas will, at the moment of their initial regeneration, have identical memories and personalities. Subsequently, they will begin to diverge. Nonetheless, a large measure of a person's uniqueness becomes irrelevant.

One may wonder whether the associated souls reside in the data, as well as in each one of the replicated beings. Whereas the human soul is very relevant to this book, its particular location—on disk, in the body, or in the sky—is of no relevance to this chapter or this book.

Thus, it is only the scientific existence of the data, and its potential use in the reconstruction of the human (or other living entity) that is addressed here. Those whose interest in the location of the soul warrants further exploration with respect to the miracles of molenotechnology, may consult their appropriate religious texts and spiritual guides.

Returning to the earlier *Star Trek* quote, "Beam us up, Scotty," the fundamental engineering principles of a design of a *Star Trek* transport mechanism are now clear.

For simplicity, let us imagine a transport between two such devices. (The projection to a location at which there is no receiving device is an additional complication whose discussion will add nothing to the point of this chapter or book.) The elements of transport depend on several things: (1) the ability to atomically disassemble and record the identity and exact locations of all the molecules of the human body, (2) the near perfect transmission of that data (information) to the receiver, and (3) the ability to subsequently reassemble, from a stockpile of necessary elements at the receiver, the precise molecule-by-molecule human whose design was transmitted.

At the practical level, transport would occur like this. The person desiring transport enters the sending mechanism. In this mechanism the person is disassembled molecule-by-molecule, during which process the disassemblers map and store all the information of the body and brain. This information is then transmitted, using current or future communications media, to the receiving mechanism. Within the receiving mechanism, the information is disseminated to the assemblers, which then flow into the appropriate nutrient bath. The person is reassembled, the bath is drained, and the person emerges dry within seconds.

The initial (prototype) transport device might be very slow due to the slowness of disassembly and reassembly. As disassemblers and assemblers improve, the speed will increase to that shown in *Star Trek.*

The transmission media already exist.

As was discussed in the section on molenotechnology, the existence of molenorobots with brains more powerful than our own, is less than a decade away. As discussed in the previous section on rejuvenation, task specific molenorobots will be able to disassemble and reassemble things atom by atom; they will record all atomic bonds and locations as they do so, and will use stored information, respectively. Eventually, a human could be disassembled and reassembled within seconds.

There are many imaginable extrapolations of these processes. For example, almost all a person's information could be stored previously (as discussed in the *Rejuvenation* section). Perhaps only an update on the mind data would be needed. But, then why disassemble the person at all? And then the person arrives where they are being sent, and the person is still where they were to depart from.

And too, since the information can be broadcast anywhere, any number of copies of the person could be simultaneously sent to a variety of places or all to one place. They would all start with the same being, but they would begin to diverge as their new subsequent experiences began to shape their minds individually.

It takes only a small extension of one's thinking to realize that once a given individual has been disassembled (for transport), and all the relevant data stored for transmission, that data could be held forever, simultaneously transported to any number of receivers, or deleted. The significance of these capabilities on the structure of society, and to our philosophical and religious views of our existence, importance, and purpose will be profound.

In the specific imagined mechanism for transport, there will be a momentary storage delay of the data received from the disassemblers before transmission. (This delay would not be absolutely necessary, but its possibility serves an important point, and is consistent with the storing of such data as was considered in the *Resurrection* section.) If we imagine for a moment all the data for an individual held in data storage, for that moment the individual will exist as data only, neither alive nor dead. In this digital form, correction or editorial changes could easily be implemented as suggested in the *Resurrection* section, for better or worse, or the whole person could be deleted by an anonymous viral progenitor.

Murder

The foregoing developments will certainly cause a reassessment of the value of human or any life.

What is the meaning of murder if the victim has been mapped and may be regenerated? How serious is the offense when the victim is not really dead, since they may be regenerated in the next moment? Does then murder become only mayhem? What is the meaning of mass murder?

Perhaps the determined murderer will have to kill the person and erase the disk. How will such crimes be viewed? The first act, the murder of the immediate person, will be much less severe since the person can be regenerated, even at a hospital a thousand miles away. But then, too, the erasure of the disk is little more than a videogame.

Danger

With all the blessings of technology throughout the millennia, it has also had several outstanding negative impacts, one of which is the capacity of a single individual, devoid of official political or military power, to harm a large portion of society.

Possibly the most immediate danger of evolving molenotechnology is its potential use to the terrorist and insane.

In ancient times perhaps the most significant destructive act by several nonenfranchised individuals against the rest of society occurred in Alexandria, Egypt, where the Serapeum and its collection of some 42,800 papyri were presumably destroyed in 391 A.D.

Some twenty years ago a news article reported that a 14 year-old boy had designed a nuclear explosive device. Experts reviewed the plans and concluded that it would have a high likelihood of detonating if it were constructed. Certainly, terrorists already have these weapons.

The Aum Shinrikyo's March 1995 use of sarin to poison passengers on a Japanese subway, killing twelve of them, was only the tip of the iceberg of their stockpile of deadly weapons and intentions, backed by a treasury of $1 billion. The Oklahoma City bombing is another example of the expanding power of one or a few individuals to harm large numbers of others.

The blossoming flow of information and availability of technology throws increasing destructive power (digital, nuclear, viral, poisonous, etc.) into the hands of single (disadvantaged, frustrated, nonenfranchised, or disturbed) individuals. The danger of the rifle (automatic or not) will soon become as irrelevant as the club (still dangerous but little used), and it will then be seen that our attention to it now is a wasteful drain, since far more threatening issues loom.

How many of us would be willing to see the powers of God in the hands of a child?

The tools of molenotechnology will not even be safe in the hands of the family next door, for a minor oversight in the John Doe family will leave at risk not just the household, not just the neighborhood, not just the city.

For even the most well-intentioned person to have complete access to the powers of this advanced technology would expose the whole planet to unacceptable risk. Even the most enlightened of us will only poorly comprehend the potential impact of whatever we do with this technology. It will not be like playing with matches, where we may burn our hand, the barn, or a field. No. If with matches we may burn our fingers, then by analogy with molenotechnology we may decimate the planet overnight.

Neither a nuclear winter nor a century long darkened sky, pregnant with ash from a gigantic meteor impact, will provide either a comparison or a metaphor for the gray goo that could cover the planet in several days after the accidental emergence of a particular molenotechnology replicator.

Against all foreseen dangers there will be safeguards. On the other hand, even if an unforeseen danger were to never appear, safeguards all have weaknesses, and none offer more than high probabilities of protection. For all safeguards there is some probability for failure within any given moment. Thus, no matter how miniscule the probability of failure per unit time, as time passes the probability for a failure within the total elapsed time grows. Eventually a failure occurs, as it did at Chernobyl.

As we take this quick gigantic step closer to the civilization of the Krels in the *Forbidden Planet,* where one has but barely to think (or dream) of something to make it real, we must ask ourselves if our security has become irrelevant, and whether the pleasure of our immediate moments has erased our interest in long-term survival.

If these molenotechnology tools are so powerful that innocence can inadvertently guide them to sucking all of us and Earth into oblivion, one must ask if such tools can be safely put in anyone's hands.

Regardless of all the direct and inadvertent dangers implicit in molenotechnology, there is another question that stands beneath, behind, before, and above all others: Which humans are trustworthy enough to have god-like powers while the rest of us have none?

The way we choose to reduce the direct and inadvertent dangers of molenotechnology may be necessary, but it may also prove to be ultimately more dangerous than the threat. This dichotomy is due to the immediacy of one threat with respect to another: the less immediate threat tends to be viewed as less threatening, even if it is more surely mortal in the long run.

The desire for security will drive us in directions that will ensure our vulnerability to an abusive leadership, who may ultimately prove to be more dangerous to us than the immediate dangers of accidents and terrorists. (See the *Freedoms Willingly*

Surrendered section.) One cause will be our habitual willingness to abdicate responsibility for our safety to the leaders, who are always ready and expected to promise it.

This may be partly out of our laziness and partly out of perceived necessity, but in the future this abdication will be a more dangerous avenue than it ever has been before, since the leadership's destructive power at the individual and global level, and their precision of control over this power, will be greater than it ever has been before. The combination of the increasing technological, global, and precision destructive capability, together with the increasing anonymity of the culprit who presses the key, further pushes this mortal issue to the brink. (See the *New Battlegrounds* section for a discussion of the evolving weapons and possible weapons that will become available to societies, corporations, and individuals.)

Furthermore, whatever tools and benefits we willingly and legally deny ourselves (the average citizen) for the benefit of public safety, will still be available to determined terrorists and others whose needs are dark.

In conjunction with the foregoing risks, the molenotechnology achievement of immortality and other medical miracles, despite their exciting allure, will increase the danger.

A future of immortality is not idle speculation; neither is it science fiction, nor fantasy. It is emerging quickly, now. Anticipation of immortality will create conflict.

The initial waves of belief in immortality will encounter some religious resistance and denial; ethical and philosophical issues will be raised and argued. Some of this oral upheaval will simply be a manifestation of the natural emotional disturbance with which Homo sapiens usually greets change. Even as the technical feasibility for immortality is achieved, there will be continuing resistance, some of which will be due to envy caused by the disparity between those who can afford it and those who cannot.

As with any new technology, it will be expensive at first. Thus, initially immortality, though it will be promised for all, will be available only to a few—those with access through wealth and power. For some years only the ruling class will be able to afford it. The rapidity with which immortality becomes globally available, offering the richest of all gifts to all humans, will be a critical issue.

Would it be fair to be disparaging of the lucky ruling class for their earlier accession to medical miracles? By analogy with today's world we average citizens of the wealthy nations serve ourselves before we worry about the Third World countries. In regard to the emergence of immortality, with respect to the ruling class we average citizens will analogously be in the position of the people of those lesser countries—helpless and exposed. Despite our goodwill, if we wish to survive, we will need to separate the hope and promise from the reality imposed by those who will be in control.

Ten years is a conservative estimate for the time until a critical juncture is reached regarding immortality. Achievement of the technology of immortality will be complete within twenty years by a conservative estimate, but this is all that is necessary for an approach to an abyss within ten years. It is only necessary that ten years from now sufficiently healthy members of the ruling class believe that within ten years after that time (twenty years from now) immortality will be achievable for them. For once they believe it, they will assume it is theirs, and they will take possession of their own immortality, as surely as we have assumed possession of Earth's waters even before we have used them all. And they will fight any threat to it with religious fervor and righteous determination, for they will indeed be fighting for their lives.

They will say and do whatever they must to survive, just as we have always done. And to whatever extent they still need us as workers to bring this miracle of immortality to fruition, they will woo us. They will speak of immortality for everyone, but in their hearts they will only be thinking of their own, and solely toward that end they will press.

It is well to remember that in ancient Egypt, commoners were not allowed in the tremendous temples, which are still so awe inspiring today, nor could they afford tombs. It was only a small percent of the population who could achieve the immortality promised by having a tomb and future sustenance delivered regularly. To this end, the average person was enslaved and sacrificed.

Thus will emerge a ruling class of immortal beings with godlike powers of creation and destruction reserved almost exclusively for their use and enjoyment.

Who will we be to them? How will they view us then?

2

FUTURE TECHNOLOGICAL DEPENDENCE

Fire, Air, Water, and Earth

Sir Francis Crick, a joint 1962 Nobel laureate in physiology for his part in unraveling the secrets of the DNA molecule, in a colloquium for the Physics Department at the University of California in La Jolla some years ago reminded his audience how special life on Earth may be, and its possible uniqueness in the cosmos.

Nonetheless, there is a progressive tendency to embrace the idea of alien life forms, and an assumption of their existence. The positive side of this tendency is its suggestion of an opening of the human mind and spirit, something humanity's conceit constantly resists.

And of course, knowing that we on Earth do exist, provides a statistical sample (scientific example) for an increasingly digitally oriented society, and thus we assume other living planets must also exist. As wonderful as such receptivity may be, the statistical truth is that Earth is one datum; one datum statistically provides zero confidence when it comes to ferreting patterns, trends, and possibilities of other life in the universe. For example, this would be like concluding the space shuttle was failure proof after successfully flying it one time.

The special existence of life on Earth, and particularly Homo sapiens' existence, is dependent on fire (the sun, wood, coal, oil, electricity, etc.), air, water, and food. This list is prioritized in order of importance to immediate survival. (Gas pressure of the atmosphere is also needed.) Without the sun and remnant (radioactively and gravitationally induced) volcanic heat of Earth, we would freeze solid in an instant; without air we would die in five minutes; without water we would die in 3 days; without food we would die in two weeks.

In pre-historic times when we lived in the wild near the equator, there was warmth, air, water, and food surrounding us, and leaders had no control over our access to it. A recluse, a wild freeman, or a runaway slave had free immediate access to all these things. (Certainly there were famines and starvation due to droughts and poor hunts, but still the leaders had no control over these mortal incidents.)

The availability of free nutriment resources (air, water, and food) has always been fundamental to freedom. This is still true today despite its apparent irrelevance to our modern Western society, in which historical imperatives encourage us to take continuing availability for granted.

For example, the potential for resisting a government that seeks to further subjugate its citizens depends on freely available nutriments. The past availability of such resources has implicitly served as a partial restraint against increased abuse by many governments. The thought of large numbers of vengeful, elusive, self-sustaining citizen-guerillas roaming their own homelands is enough to keep any government in check. This restraint is fast fading.

The Natural Environment

Homo sapiens is creating a revolution in the environment of Earth. The living planet will at worst become a dead planet, and at best become deadly to almost all currently remaining species. Possibly new organisms will evolve to enjoy the changing planet, and certainly Homo sapiens will survive. Because of advances in technology, soon a healthy natural environment will be unnecessary to Homo sapiens' material well-being.

Most of us care deeply about possible loss of the environment. Even loggers, whose livelihoods depend on logging, even the most hardened hunter, all of us at least occasionally look about us in awe and wonder at the incredible construction in the cathedral of nature. All of us in our deepest heart of hearts, in the most sensitive remote regions of our souls, in the shrinking nature-portion of our spirits, feel some pause, some uncertainty, some primal fear, that it might all disappear.

The fate of the environment rests almost solely in human hands. Whether those hands withdraw in peace, or squeeze every last drop of life from Earth, depends on the balance between the depth of our concern for the living planet and the myriad forces that encourage us to continue on our present path for its revolution.

Our Strengths are also Our Limitations

Despite our deep concern for the environment, our ability to see and our dedication to change our course are hampered by even our greatest strengths: (1) giving highest priority to what is immediate; (2) inductive reasoning; (3) propensity to ignore what we cannot see; (4) adaptability; (5) trust in our leaders; (6) desire to continue our comfortable lives; (7) miracles that technology promises.

(1) Giving higher priority and greater credence to immediate problems, than to those that are more remote, is an excellent survival strategy.

Survival depends on many choices and while no strategy is guaranteed to work in every situation, strategies develop that work best on the average. It is reasonably clear that from a survival standpoint, generally giving priority to one's immediate safety, as opposed to worrying about the future, is the the most successful strategy. And for humans, what one is unwilling to worry about is very closely linked to what one disbelieves, a clever instinct developed (through god or nature) to urge that implicit viewpoint.

In the jungle, one might lie awake the whole night worrying

about a possible encounter the next day with a lion or a cobra. Better to go to sleep, and thus be rested and better prepared to meet tomorrow's challenges in the morning. The success of this strategy gradually imprinted itself in the DNA of the whole gene pool, and ultimately became an instinctive behavioral mode.

On the other hand, through our intellect we learned to plan ahead, especially as we migrated to regions where planning was important to survival. This would have been the case as we moved further from the relative comfort and abundance of the tropical jungle. Nonetheless, our tendency is still to give our immediate concerns higher priority, and immediate threats greater credence, because that is instinct, which we follow until we have good reason to the contrary. It is likely that planning began to take precedence over instinct only on a case by case basis. Civilizations and their constituent individuals probably learned by hard examples how and when to ignore this useful instinct. And too, this is why it is so difficult to educate children. The feeling most of them begin with, and often carry into their adult years, is that if one has snacks and friends nearby what possible use is an education?

Because of the strength of human imagination, to believe some threat lies in the future, to give it credence, is to give it immediacy: One is motivated to act now, to prepare for the perceived threat.

This is the binding and blinding relationship between credence and immediacy, and it works also in reverse: if the proposed threat lies too far in the future, then it is given less credence. That still is our tendency, and it strongly influences our perception of potential problems and their solutions, and thus guides our behavior, unless we have clear examples to make us feel differently, or until the future draws near.

For example, it is popular today to think of electric cars as environmentally preferable. Actually, solar cells and batteries are very toxic to produce, as are other portions of our technology that make electric cars possible. To create a complete comparison for environmental purposes, regardless of the energy storage technique used, one must consider the environmental impact of the production of the car, its use, the energy source, and the amortized pollution inherent in bringing the civilization as a whole to the technological position of being able to produce a technologically advanced car. Perhaps there is a study that shows that the total environmental pollution of an electric car is less than that of a gasoline car; perhaps not. But this is not what most people care about. For example, the leaders and people of Los Angeles only care that there is less pollution in the skies above their city, in the air they breathe now, and for that purpose, electric cars are preferable, regardless of their global impact.

Similarly, individually most of us will choose a solution that decreases the pollution density in our homes now regardless of whether it increases the global average pollution density and increases the total global volume and mass of toxic waste. Such choices seem appropriate, justifiable, and excellent since they move the problem to a space in which it is less noticeable now and can be tolerated until a later time, when other people (or beings) will have to deal with it.

(2) One of Homo sapiens most potent survival tools has been its capacity for inductive reasoning, by which a general conclusion is drawn from particular instances.

This tool probably has its roots in an inductive survival instinct: for example, even without much intellectual capacity, an animal will (learn to) look for water where it has found it before.

From the earliest glimmerings of human civilization, recognizing similarities and repetition of events has enabled us to increase our tribal and national wealth and health, both pertinent to our security and survival. Furthermore, induction has given us the power to identify natural laws, even in the subatomic realms, and to consequently predict, design, and construct tools that promote our material well-being, and to reach beyond this planet and into the solar system. Already, with instruments we have reached outward to the edges of the cosmos and inward to the subnuclear domain of proton structure.

On the other hand, this incredible intellectual tool has a blind side. It encourages us, in a crude sense, and usually to our benefit, to assume what has been yesterday will continue tomorrow. This is akin to the weatherman predicting that the weather for tomorrow will be the same as the weather today; this usually works, but it can be terribly wrong.

Alternatively, this intellectual inductive capability is inherently related to our belief, held for so many years, that Earth's resources were so vast that they could never be depleted; likewise, we dumped toxic waste and it seemed nothing bad happened, so we did it again, and again, etc. And now some bad things are happening, or being discovered to have been happening all along.

(3) A surprising find of child developmental studies, by Jean Piaget (1896-1980) and others, was that initially a human infant has no sense of object permanence. This is the comprehension that something that is hidden still exists. This understanding is not innate in humans; rather, it develops during infancy.

In front of an infant, Piaget would hide an object under a beret, which was covered by a blanket. At about the age of 18 months an infant would seek and find the object. This is when an infant begins to comprehend the existence of something unseen, especially if it is something the child wants.

On the other hand, this developed capacity coexists with our proven continuing innate ability to ignore the existence of something uninteresting, or, beyond this, to actually disbelieve the existence of something unpleasant, especially if we have covered it up. Of course, we may intellectually know it is there; nevertheless, as adults, we continue an innate tendency to disbelieve, ignore, or give less importance to what we cannot see: "Out of sight, out of mind."

From a survival standpoint it is certainly reasonable that we are thus constructed. In the prioritization of the many tasks for living, it is usually our best strategy to give precedence to what is most obvious. The lion in front of us is more dangerous than the lion in the distant forest or than the ants beneath our feet. Unfortunately, humanity has created a vast array and huge volume of hidden dangers, many of which have an importance far surpassing more obvious dangers, such as national enemies, for example.

The constituents of these hidden dangers are in toxic wastes, garbage dumped at sea, and landfills. Toxic wastes from automobiles, factories, fertilizers, technology, etc., spew into the atmosphere, fill the sea, cover the sea floor, and enter all bodies of fresh water. Solid wastes dumped into the sea, generally add to the pollution of the sea and also create zones of death, such as the two-mile diameter one in the Atlantic Ocean near Manhattan. A sizeable fraction of the Baltic Sea floor is dead due to years of fertilizers coming down Russian rivers. Diapers, for example, constitute cubic miles of landfill.

A poignant example of our confusion about what we don't see is in biodegradable plastics. In their manufacture, plastic bags leave behind toxic by-products, as do most of our modern conveniences. Additionally, when we are through with plastic bags or bottles, they become an eyesore.

Eventually, motivated by environmentalism, the idea arose to make plastic bags disappear through biodegradability. This idea, or at least its result, may have been misguided. First, the question has arisen as to whether biodegradability works, in other words, whether the items biodegrade. Second, a more disturbing question arises:

> "Even if biodegradables work, what do they degrade into?" asks Jim Middaugh, a spokesman for the Environmental Defense Fund. "The term 'biodegradable plastic' is a misnomer. Essentially, what you leave behind is certainly less visible, but certainly more dangerous, plastic dust." While the point is arguable, Middaugh maintains that plastics may contain toxic additives, which could be more harmful in a dust form than in a bag or a bottle left lying intact on a roadside (Stuller).

Regardless of the microscopic or macroscopic size of individual pieces of waste, their incredible aggregate volume has surpassed what was once viewed as Earth's infinite capacity to absorb. Thus, we are beginning to see some of it, and what we do not yet see directly is having effects that kill.

(4) Another of our powerful tools for survival is our adaptability and habituation to change.

An often heard environmental appeal is that "we need to preserve our precious environmental heritage for our children to enjoy." This suggestion has a hollow ring to it. Humanity has shown an aptitude for enjoying many different kinds of things. We and our children can and will learn to enjoy other things, such as videogames. It is ironic that the manufacture of these things pollutes the environment, which we can gradually learn to enjoy less.

Environmental attitude is, and has been, molded slowly. If in 1700 a technologically advanced alien force had invaded Earth, and done to it in one year what we have done to it since then, we would have been horrified at the destruction. Terror would have filled our hearts, and it would have turned to rage. To the last man, woman, and child we would have fought these destructive aliens, despite their "need" for "our" planet. Most of us don't feel the rage because the destruction has taken place according to our choice and relatively slowly.

What is happening with the environment and our response to it is a simple habituation to change. Some of us want the environment to remain as it is, because we have learned to enjoy it that way, and change is uncomfortable. But our children have already come to expect less of it, and will learn to appreciate there being less of it. The rate at which the revolution proceeds is a balance between our immediate appetite for more material benefits, and our necessary habituation time to the implied environmental changes.

(5) Trust in our leaders is an essential ingredient in successful organizational behavior.

Almost all social systems and nations have relied on some form of leadership for the successful conduct of their existence, and for guidance to successful survival. Our trust in such guidance is probably necessary for stable social and government systems.

Directly or implicitly we expect our leaders to guide us, to fulfill our perceived needs, and to encourage a point of view that allows us to accept various incongruities and rationalizations that remove emotional and psychological barriers to what is viewed as our reasonable and rightful forward progress in the conduct of our individual lives and the advancement of our civilization.

A stunning example of point of view, and relating again to hidden dangers, is in the presentation of results of The National Acid Precipitation Assessment Program (NAPAP), which concluded in 1991: "The most celebrated of the NAPAP highlights' good-news conclusions was that only five percent of the 4,330 New England lakes sampled are acidic" (Moore). This number, five percent, seems like a small number compared to one hundred; however, it is an enormous number compared to zero. The use of the word "only" is a critical pacifier for the average reader. Furthermore, the definition of "acidic" may worry only the discerning reader. By the NAPAP definition, at that time a lake could be too poisonous to support rainbow trout, for example, and still not be classified as acidic. Numbers and words can be used to suggest almost anything. This results presentation was meant to assure us that no problem exists, and for many of us that was comforting news; however, an environmentalist could have been horrified at what was found.

In addition to government and religious leadership, it is the place of the corporations, who employ us, to provide a focus for our hopes and dreams of future well-being, and to set standards for our expectations and behavior. In this they provide a path to fulfillment and satisfaction, and a unifying tradition of purpose, wants, and needs. They do this through their powerful, mind molding, advertisements, so artfully created to tap into our insecurities, envy, and status fears, and to push all our peer pressure buttons, so that we will consume more of what they sell.

Population and consumption support successful businesses, create jobs, and are causing a revolution in the planet's environment. Simultaneously, we seek reassurance that the environment will be taken care of. Thankfully, the big corporations are willing to do the job and to inform us of their fine progress.

Chevron Corporation sets a good example in their beautiful "People Do" print advertisements. These environmental advertisements tell of various beneficent acts of humans toward nature, and subsequently always end with a wonderful leading question about whether people care for the environment and other species, and a true answer, "People Do." These give most of us the reassurance that we need, that we do care about nature and the environment, and that the threats to it are being squelched and the problems are being fixed. (See further discussion about this kind of verbal smoothing in the *Religion, Morality, and Esteem* section of the *Sacrifice* chapter. See also thoughts about the momentum sustainment offered by the powerful concept of "People Do" under *The Balance* section.)

While stabilization and unification are important to organizational survival, the methods sometimes, perhaps too often in the present world situation, lead us astray. This again relates to the issue of what is hidden and what we see. For example:

> In a remarkably candid interview last fall, Mike Levy, a Washington lobbyist for Mobil Chemical Company, admitted that the degradability claim on Hefty plastic bags "is just a marketing tool." Mobil developed the bags because marketing studies showed that consumers wanted them, and competitors had such products in stores, Levy told a reporter for the *Marin Independent Journal*, a daily newspaper in northern California (Stuller).

Our trust in leaders is in some ways like that of children who must depend upon their parents, right or wrong. Our leaders have been essential in our "progress," in getting us to where we are now, whether or not our position can be envied.

It is natural to have trust in those on whom we depend for clear vision, and who have a talent for and natural tendency to smooth difficult circumstances with honeyed words, so that we may more easily follow our startling path into a radically changing future.

(6) We have an understandable desire to continue our comfortable lives, or materially increase them.

This is motivated by our basic survival instinct and intelligence, which suggest that more material well-being now is better for our survival. Stated in a different way, increased material well-being gives us an instinctive feeling of security. More basically, we feel we won't starve before the cupboard is bare, so a well stocked cupboard gives us a good feeling, a secure feeling.

Most of us belong to or have families, and along with our concern for our individual selves is our deep-rooted concern to see our family members live well, be educated, and own a home. And we (at least in the developed countries) have long felt and been taught that these are basic (human) rights in our existence.

These feelings, beliefs, and associated habits of existence either will prevent us from seeing, or will allow us to see only too late, their implied near-term danger.

(See related material in the section *The Power of Pleasure* in the chapter *Blindness*.)

(7) The attraction of the miracles that advancing technology promises us in the near future—the material well-being we can have as well as the irrelevance of the environment to our having it—is a powerful motivator for our continuation on the present course.

The appeal to people that they need the living planet in order for themselves to survive has always had a heartless ring to it. For example, there is a threatening claim that if we burn all the rain forests, pollute the land, and/or pollute the oceans beyond the plankton's ability to survive, then there will be no oxygen. The cry is that without the living Earth, Homo sapiens cannot survive.

This claim is utterly false.

A profound result of molenotechnology is that eventually humanity will not need the environment of the living planet in order to survive. Through molenotechnology and related advances in other technologies, Homo sapiens is developing tools to literally get oxygen, water, and food from rocks. All the appropriate atoms are there. The only missing tools are atomic robots that can work on an atomic scale. Molenotechnology will provide them.

Technology will be able to provide everything needed, regardless of the complete toxification and destruction of the planet.

If people do not care on behalf of others—the other species—then technology has always offered a covert saving promise. And now, in molenotechnology, people will see that promise is coming true, and they will assume it will be true in time. Motivation for preservation of the planet is lessening already, and will continue to decline more rapidly as the promises of molenotechnology loom nearer and larger.

Thus, Earth's atmospheric oxygen, ozonosphere, water, plants, and other living things will soon be irrelevant to the existence of Homo sapiens, and our concern for these will fade.

This is happy news for those who are satisfied with only man's survival and preeminence, regardless of the cost in Earth life. On the other hand, it is triply bad news for the most stalwart environmentalists: (1) they lose the living planet; (2) they lose the implicit satisfaction that would have come with the whole planet living or dying together; (3) they will die despite the survival of Homo sapiens.

The dichotomy in this third statement is due to the disparity between (1) the rate at which advanced technology production will supply the general population with artificial environments to survive on a dead planet and (2) the rate at which population, consumption, and toxic technological by-products are killing the planet so that it will be unable to support any of us naturally.

Paradoxically, one of the exciting promises of molenotechnology is environmental cleanup.

One way to proceed would be this. An analysis would be done of kinds and quantities of the various toxic substances (and other waste) that Homo sapiens has produced and deposited, dumped, spilled, or otherwise let slip through our global fingers. For each of these materials, specific molenorobots would be designed and constructed (in molenorobot factories). (We will assume that there are no production waste products to concern us.)

Subsequently, these molenorobots would be set loose to replicate and to gather, disassemble, and neutralize absolutely every single toxic molecule that man has ever produced! This would include the two-mile diameter death zone of human waste in the ocean off of Manhattan. The by-products would be predetermined safe molecules, such as water, etc., carbon dioxide (too much of which we already know to be unsafe), or molecules of elements such as carbon, oxygen, argon, nitrogen, etc. And at completion of their task, the molenorobots perhaps would disassemble themselves into atoms and safe molecules.

This is not fantasy. It will indeed be possible! On the other hand, the real issue is whether we will choose to use the new tools this way. It seems unlikely that Homo sapiens would squander its new-found technological and production capabilities on environmental cleanup, which would take material, time, and effort. In the future, as in the past, we will have other needs to be met. And as previously mentioned, our needs have always expanded to keep pace with whatever material benefits we could have.

For example, consider this. The knowledge and technology for how to not pollute, not poison, and not injure the planet has existed for a long time: Simply don't do it.

In the 1800s there were people giving environmental warnings. The January 1872 *Scientific American* commented as follows:

> The habits of the present generation are such as to give rise to more refuse matter and poisonous products than those of previous ages. The fuel we use, the articles we manufacture, and the waste of sewage continue to create more impurities than were known to our forefathers.... The true remedy is to stop filling the sewers with matter that no power can afterward cleanse.

People simply could have chosen to not do what they did: they could have chosen to have fewer children, so as to reduce rather than increase the size of Earth's consuming population, and to have less rather than more material goods.

In the Garden of Eden there was abundance; Homo sapiens wanted more. At any other time, say in the early 1900s, there was abundance; Homo sapiens wanted more. Today there is abundance; we want more. But, of course, the veracity of these comments is just as relative as is abundance. We have the potent attribute of habituating to our abundance as well as to our losses. And then we want more again. Someone is always aggressive enough to step forward to address our increasing expectations, and to attempt to lead us to their fulfillment.

The implication that we would conduct an environmental cleanup is a fantasy that overlooks man's timeworn habit of needing more in conjunction with his ability to take more. On the other hand, there will surely be some window dressing activities, just as there are now, to appease our somewhat troubled souls. These environmental molenotechnology projects will be analogous to California's redwood highway, where hundreds of trees were left standing alongside the road so people could get the feeling of the vast redwood forests that used to be there. That feeling was sufficient for the average citizen of the past and present, and doubtless it will be sufficient in the future.

More to the point, however, is that in the future Homo sapiens' concept of garbage, pollution, and destruction will radically change. Through molenotechnology we will have the capability to literally process the whole planet, from its crust to its core. With absolutely every atom within the potential grasp of Homo sapiens, the central issue is that we have that grasp, and what we do with it.

For example, medical attention is always high maintenance: it is expensive and environmentally very destructive. People who are sick, if they have a choice, will want the technology turned toward their immediate problems, not toward environmental cleanup. This will become no less true with the potential for immortality. The trade-off between our immediate desires and the environment will continue to be made in favor of our immediate benefit as it always has, as long as we are convinced that any possible bill for the trade is very remote.

To imagine that there will be no waste products from the combination of goods that would please humanity with such awesome production power is far fetched. Presumably, production could be more efficiently balanced in a way that will reduce the amount of waste associated with a product, but there will still be waste.

For the ruling class, who will have increasing power to have the promises of the technology fulfilling their lives, the amount of pollution and waste is likely to increase along with their vastly and rapidly increasing material benefits. If it doesn't increase for the average citizen, that will be because the average citizen is not getting an increase in his material standard of living.

Also, processing of the entire planet, turning it into some futuristic material vision, makes past meanings of garbage irrelevant; if what we presently perceive as garbage were to disappear, the danger lies in what will replace it.

Garbage will take on a new meaning. In a sense, with regard to the health of Earth, most of what we produce will be garbage, but we will love much of it, initially. For example, the way most of us think, it would be heaven on Earth if the whole planet were turned into gigantic homes with chemically treated swimming pools, Cadillacs, and videogames. Perhaps too there would be zoos for a few representative animals, but they too will have to be enclosed in artificial environments to keep out the poisonous atmosphere, etc. (And we will commend ourselves for rescuing them from death on the outside.)

Among the other falsehoods to be associated with evolving molenotechnology are that there will be sufficient safeguards against some of the foreseen dangers (but not the one presented in this book), some of which would make the environment meaningless. (See the gray goo idea under the section *New Nondigital Weapons.*) As previously mentioned, there will be safeguards, to be sure, and the ruling class will guarantee them to be safe enough; the problem is that they will not be failsafe, and those of us who stand to benefit the least and last from the promises of the technology will be the ones most at risk when the safeguards occasionally fail.

It may only take one failure. To imagine the potential extent of a safeguard failure, one can extrapolate examples from past to present. In the time of cavepeople, fire was high technology. Probably getting a burned hand was a minor accident. Perhaps a major safeguard failure resulted in the accidental burning of a large expanse of forest. But even such a catastrophe was also a naturally occurring event, in whose immediate aftermath healthy life arose again, over the same terrain. Moving to modern times, one must compare the Chernobyl nuclear power plant disaster in Russia. The duration and expanse of the problem, suggest a frightening trend. Extrapolation to consequences of a safeguard failure in the future, with a technology as powerful as molenotechnology, suggests global death and destruction.

Population

Increase in population is another human strength that carries an implicit weakness. So much instinctual pressure, joy in childrearing, and production dependence encrust the tradition of increasing population, that the balance between its strength and weakness is badly obfuscated.

It is the natural instinct of all species to breed, and in this Homo sapiens is no different. This is one of the strongest elements of the survival drive. No species survives without it.

To have a baby is magic. It is a miracle. It is maternal and paternal instinct. It is motherhood. It is fatherhood. It is wonderful.

Everybody seems to benefit when a baby is born.

The mother gets lots of attention from relatives, friends, and passersby. The father's virility is proven, and his position (in the traditional family) is increased by the family's greater dependence on him. Social contact is increased through school and childhood extracurricular activities. The parents have someone to follow them around and pump up their ego by asking all kinds of impossible questions, and otherwise treating them as gods, at least in the early years; the difficult years apparently don't negate the benefits. In the past, the parents had the security of knowing someone would care for them in their old age.

And the parents just feel more normal, and acceptable, since, of course, everyone else is planning to have children, has children, or has had children. It is traditional.

Being around children who are growing, and participating in their growth, is a wonderful experience, and it is no less magical than giving birth itself. To see children, or the young of any species, become coordinated, learn, and to focus to accomplish things is as miraculous as anything could be.

Also, having children is one of society's expectations. Having a baby has usually been perceived as making a contribution to society. Both the state and church in western society have encouraged this.

Business leaders and average persons always seem to like saying and hearing, "Our community is growing." This is ever on the lips of real estate brokers. There is pride and security in numbers. Even when people know that the population increase of their community will imply the drinking dry of streams and lowering of water tables, and the general obliteration of natural beauty as more of it is converted to make room for the welcome "increase," they have a sense that all is well, and indeed improving.

Property values increase as demand increases. All original owners feel more secure as their property values rise with the increase of population.

From a business standpoint, the growth of a community often means that local businesses will flourish more easily, and that the city budget will increase through increased taxes. Everyone feels more important and safer when population is on the increase.

National leaders and the ruling class in general recognize that population increase guarantees a steady supply of farmers, factory workers, and soldiers. Churches love the increase of their flocks as much as do sheep ranchers. The biblical quote, "Go forth and multiply," is an age-old imperative for all Homo sapiens.

At every level of society the goal of increasing population is still promoted and encouraged, and research scientists look for ways to promote fertility so that infertile women or impotent men can have children.

Increase in property value is in large measure due to increasing population, rather than reflecting a change in intrinsic worth. Nonetheless, it makes us feel good to see our house value increase, and this implicitly promotes increasing population.

How many of us ever wonder about the extent to which our enjoyment of the increase in our property value is derived from the amorphous aggregate of all children wanting what we have? Is it so different than the enjoyment we get from having food for them, knowing that they will get some soon? There is some instinct that promotes our enjoyment in seeing others with less than ourselves, or at least in seeing our resulting apparent relative net worth increase. That each of us contributes a few children to the growing aggregate, maintains a population of the less fortunate without being personal about it. (From a standpoint of maturing there is of course no reason that less experienced younger persons should have all the privileges of those with more training and experience, but that is another matter.) It just makes us feel good.

How many business people would be bold enough to step forward to make their businesses run in a declining population? Obviously some, possibly many, businesses would go bankrupt; almost all would get smaller. We have been taught that if your business is getting smaller then you are inadequate. The veracity of this attitude is seldom, if ever, questioned.

How many people consider exactly how many animal deaths are guaranteed by having a baby, and that having a baby contributes to the genocide of another species, and contributes to the death of another portion of the planet?

Why would anyone allow these ugly truths to spoil the magic of birth, and the excitement of having children?

Nature has the capacity to convert carbon dioxide into oxygen through plant photosynthesis. Homo sapiens and other animal species need oxygen to survive, and they give off carbon dioxide when they exhale. The global symbiotic balance between plant and animal has been a wonderful miracle.

An exemplary, simple, hard, categorical truth on Earth today is that the present human population and its associated activity is creating carbon dioxide faster than nature can convert this gas can back into oxygen, and through deforestation and pollution of earth, sea, and sky, humans are further destroying nature's capacity to do this conversion. There is a loudly ringing bell here. How many people choose to hear it? Do we care? The threshold has already been passed.

Of those people who can hear the warning bell, for many of them the appropriate assumption will be that since the threshold has been passed and we are still breathing, everything is fine. (This is our often useful inductive reasoning at work.) Other people will recognize something must be done, and they will look to the government, or some other corporal power to fix the problem. Others will wait for divine intervention. Perhaps some will hope for alien guidance. Very few people expect to have to actually give up anything and retreat back to the other side of the threshold.

People Care

People do care, care deeply, and are very concerned about the changes that are occurring in the environment.

A logger is not simply a mechanical extensions of his screaming 13,000 rpm twenty pound chain saw. Loggers too have souls. Despite their need to make a living, many of them recognize a need for change. Dale Page has been cutting trees for 44 years in the Pacific Northwest.

> "It doesn't take long," he says, "To think it's been growing for 200 years or better, and then it's down in a minute and a half. It's kind of sad. It affects you. I don't think you'd be human if it didn't." He adds: "An old-growth forest is unique. There's just something about a big tree that makes you feel kind of small." Now he recognizes the need to protect nature from man. "We've only got this one old earth, and we better take care of it. I most certainly do not think 'environmentalist' is a dirty word. Anybody who isn't one has his head in the sand" (Gup).

J.A. Hunter, a former big-game hunter in Africa, also recognizes that something may be amiss. In an article describing his career he wrote, "In my career I killed more than 1,500 African big-game animals. I don't say this with pride." In the conclusion of his article he asks:

> Was it worth killing these strange and marvelous animals just to clear a few more acres for a people that are ever on the increase? I do not know. But I know this. The time will come when there is no more land to clear. What will be done then (Hunter)?

We may ask ourselves if he wonders, as do many of us, whether there might be some value in finding the answer and applying it now, rather than waiting until everything is dead.

We don't want everything to die, and we want more material well-being, or at least no decrease in what we already have. We know that ozone depletion and other consequences of human activity are exterminating hundreds of species per day. And people fear the potential greenhouse consequences of increasing carbon dioxide in the atmosphere, which we also know is caused by human population, standard of living, and technology. (These are examples of our environmental concerns.)

Nowhere are such concerns more beautifully, artfully, and profoundly expressed than in the Chevron Corporation "People Do" advertisements (previously mentioned). These environmental advertisements gently acknowledge the terrible conflict we face between our needs and the needs of the rest of living things, and soothingly assure us that we do care, that we are good, and that the problem will be resolved fairly and to the benefit of all living things.

As mentioned earlier, these corporate advertisements illustrate our need for reassurance and emotional comfort through difficult times, times of revolutionary environmental change. And the corporations are adept at identifying what we really want, whether it is reassurance or substance, and providing it to maintain their profit margin.

We have the posturing, rituals, and reassurance of caring, but we may not have enough substance to halt the destructive processes that we have set in motion.

The Balance

Sometimes in our eagerness to have more or to extol ourselves, and not wanting to so directly confront our grasping nature, we confuse the boundary between ourselves, our wants, our needs, and the rest of the world.

For example, one of Chevron's "People Do" advertisements offers this: "But the needs of a burgeoning world sometimes conflict with the necessity for preservation of nature's habitats." This phrase gently converts Homo sapiens greed to need, gloriously unites our increasing aggregate being with the growth of a living Earth, simply ignores the past and potential fullness and omnipresence of nature, and blithely shifts the focus to confined habitats, thus, smoothly obfuscating the distinction between who wants more and who will be sacrificed to provide it. The advertisement's suggested solution, "a balance between the two," contains a ring of justice and fairness similar to what we may have learned at our grandmother's knee. On the other hand, history has taught us that "balance" with an increasing aggressor is seldom more than temporary appeasement, to later be violated again.

The bold faced "People Do" conclusion of the Chevron environmental advertisements encourages, praises, and affirms the human bottom line that we are in motion and will continue to "do." The instinctive excitement humans feel with activity and the traditions of the Calvinist work ethic will continue to provide motivation for what has for many years been called progress and viewed positively.

Despite our proven deep concern for nature's predicament, most of us are not sufficiently motivated to make the significant material sacrifice (and it would be mind boggling) necessary to stopping detrimental environmental changes. On the other hand, most of us have enough sensitivity to feel that something is changing in a way that feels odd, or even painful. Our instinctive habits and need for comfort will urge us to go ahead and behave, with our birthing and consumption, in such a manner that the revolutionary environmental changes will occur, but we just don't want it to hurt.

That is where corporations like Chevron will continue to help us. They understand our concern and our pain, and how to guide us gently through the difficult trade-offs that will be made.

Their environmental messages provide a sweet story to make us more comfortable in the course of the environmental revolution, by obscuring the frightening facts that the planet is dying and that when it comes to making the choice, for most of us the planet's fate is irrelevant, as long as our dream life is fulfilled or promised in return.

> Mike Levy (the Mobil Chemical Company lobbyist mentioned above) said "We're talking out of both sides of our mouths. I don't think the average consumer knows what degradability means. Customers don't care if it solves the solid-waste problem. It makes them feel good" (Stuller).

Another example of environmental concern, the simultaneous need to feel good about the change, and adapting to the loss of Earth, is in the global acceptance of Earth Day, which is supported by individuals, governments, and corporations. The overt intention is good. Maybe we get in the car, drive to the local Earth Day site, and mill around in a crowd, hear a passionate scientifically earnest speech, learn a few things about recycling, have a Pepsi, put the can in the recycle bin, etc. Maybe when we get home we begin a recycling effort. It's been a fun day, we've made a change, and we feel good.

Unfortunately, our environmental efforts and plans are sorrowfully miniscule compared to what we would have to do to halt the extermination of hundreds of species per day that we cause because of our population and standard of living. People are beginning to recognize the true size of the problem, but we still may ask whether enough people see deeply enough, or want to do enough sufficiently soon, to stop the radical changes and get us off the death track.

We do care, but we may not care enough. After all, how many of us are willing to stop procreating and unplug from our technology so that, for example, carbon dioxide levels will immediately stop increasing, indeed, so they will decrease? How many of us even comprehend how the enjoyment of a technological benefit we wish for today may imply planetary death in ten years? At the moment, in the face of our family's immediate needs, that time seems very far off, and that negative consequence seems exceedingly remote. It is just hard to accept that we truly are taking it all.

In September 1513, from a mountain top on the Isthmus of Panama, Vasco de Balboa was the first European to see the Pacific Ocean. When he arrived on the beach fours days later, he entered the water and claimed the ocean and all its shore for Spain! This possessive acquisitive Homo sapiens trait repeatedly reveals itself in our culture. The above quoted Chevron "needs" advertisement provides a clear modern manifestation of it. The use of expressions such as "our natural resources," is another. Everything to which we apply this special word "our" has already been implicitly possessed by us.

The word "our" itself has so many meanings—from home and stewardship, to possession and domination, to rights for butchering—that its frequent use, for example, in the expression "our world," must offer comfort to anyone who wishes to hear any particular thing while not discerning much of anything. Its carefully chosen use can simultaneously encourage human satisfaction on all levels without its directly threatening the object of our attention.

Another example from our culture is the beautiful song written by Michael Jackson and Lionel Richie as part the global hunger rally organized by Bob Geldof in 1985, "We are the world, We are the children..." (Stambler). These poetic words imply such hope, wholeness, and unity as to fill one's heart with joy, and the melody adds to the suggestion of global harmony and peace. Even though a figurative meaning may lie behind these words, they suggest, and in their beauty lend further acceptance of, a more literal meaning. If we feel good about Homo sapiens' young claiming they "are the world," then together we shall surely destroy it.

But no one should be shocked to think that we will actually continue behaving in a manner that kills the planet. Despite a small minority of weakly dissenting voices, as an aggregate being, Homo sapiens has never shown a serious interest in saving the planet other than as it might affect its own welfare. How can we be faulted for this? How could it be otherwise? This is how successful species behave; if they do not put their own immediate welfare before all other things and beings, another more aggressive species comes along who will do so, and it overcomes the less aggressive one. This is jungle law.

To return to an analogy with children, and thinking of the aggregate of our species as one being, our deeply expressed concern for the environment may be akin to a child who wants both the ice cream cone and the lollipop, and having had to make a choice, licks the one while still pitifully crying out for the other.

Our desire to increase our population, and to have more material benefits, has been and will continue to guide our course, which is contrary to the benefit of the environment and the health of the planet. We have as much difficulty seeing the truth of this as we do in not sensing the humiliating absurdity of environmental catalogues offering doormats with the images of animals on which we should wipe our feet! Imagine the attitude of the NAACP, or any other racial organizations, if any company should offer a similar doormat with human faces on it.

On a darker note, from another perspective there is an additional force pushing the balance away from environmental favor. In the innocence of childhood many of us pulled the cat's tail. We were curious, and we were excited by kitty's reaction. On a deeper, more primitive, level, we enjoyed the exercise of our manipulative (grabbing) capacity and our superior power over another being. This instinctive tendency is obvious in play of the young of most social species, including primates, where strength and dominance are the essential indicators of subsequent leadership and pack pecking order.

While in this feline instance the primitive portion of our persona was eventually overridden by parental admonishment regarding courtesy toward the family cat, the general proclivity remains embedded in our deepest cerebral being. And as we matured toward adulthood, our conception of existence expanded beyond the family home to include Earth.

An innate feeling in most of us, originating in childhood, is that Earth is a big place, an inhibiting entity, a dangerous living being. The exercise and demonstration of our aggregate power to affect it, even to hurt it, offers reassurance that we have nothing to fear from it, excitement that we can overwhelm it, and the pleasant associated feeling that we individually and as a species are superior to it.

In our relative innocence, the current publicity regarding the environment is not much more than a ritual for focussing our attention on the death process of it. And even as a child might agonize over the death of the cat if his torments accidentally took an unfortunate mortal turn, we too feel some of that agony and remorse.

However, in this ambivalent circumstance, the thrill of the current death process is an additional motivator for our chosen course in the "balance" (of the Chevron advertisement). There is tremendous excitement about Earth life hanging by a thread and the scissors in our hand; there is a joy in the genocide we "do," but it isn't permissible to admit it.

The politically correct outrage we feel at this suggestion imperils us because it prevents us from comprehending our true nature, and hence blinds us to what others who are just like us might do to us. Thus, we have been the predators of sheep and enjoyed the slaughter; in our denial of it, we are less apt to comprehend that we are becoming the sheep.

Environmental Conclusion

Despite our obvious deep grief at the imminent loss, based on understandable human impulses for increase in population and material well-being, and the promising miracles of rapidly advancing technology, the present trends support a definite conclusion: the living Earth is dying, and it will be incapable of naturally supporting human life and will largely be dead within ten years. Nonetheless, Homo sapiens will survive.

Some people may wish to condemn or cast judgement on Homo sapiens for its behavior, which may seem selfish or reckless, but that is not the intent of this book.

The importance of now seeing beyond our grief, recognizing the reality of the future, and accepting our responsibility for presently choosing it, cannot be overestimated.

To have always had the freedom to live off the land, regardless of the small extent to which we have chosen to do so in the last several decades, has been a gift whose value will not be fully realized until we are directly confronted with the frightening significance of its absence.

There are two unsettling issues here: (1) The increasing control the governments, corporations, and other organizations will have over our lives once there is no longer any natural source of nutriments, and (2) who will decide which of us will be covered by the limited supply of artificial environments.

Thus, it is not yet clear which of us will be allowed to live.

Nutriment Production and Distribution

Until now our increasing dependence on the mechanisms of nutriment production and distribution have been reasonably voluntary, and always reversible. In the near future, we will have stepped into a new era, one in which we no longer have the choice, in which we no longer have any source of nutriments, other than those that are artificially produced and distributed by corporations and governments. Processing and distribution are two independent control points, and each will soon become a choke point.

Already production of sufficient food supplies depends upon systems that are controlled largely by entities and beings other than ourselves, as does distribution, without which sufficient food supplies are useless. We are almost all now dependent on the powers of distribution to keep from starving, which only takes two weeks. The extent of this dependence is increasing.

Gradually, all natural fresh water is being polluted and contaminated with garbage and toxic waste, as is the ocean. Increasingly, people in the industrialized countries rely on processed distributed water (liquid). Three days without water (liquid) means death.

There are many forces that continue to push us in this direction, including our own continuing willful pollution-inducing consumption. Certainly, regardless of what anyone might say, our leadership would not have it otherwise.

For example, a beverage company's profit motive is served by continual pollution. No matter how bold their pro-environmental claims may be or become, no matter what pro-environmental financial contributions they may make and tout, and no matter what pro-environmental emotional display the leaders of such industries may stage, their bottom line is profit. That is what corporate survival means. And nothing works as well as an expanding market to help a business grow.

If such leaders are challenged on this point, it is easy to imagine them assuming the posture of environmentalism and defensively claiming that they need pure water for their businesses. And that will be true. What they won't say is that the necessity of purifying (processing) the available water for their use increases their control over what we drink. Control over the available potable liquids is the imperative.

These corporate leaders would naturally prefer to have all of us sucking Coke, Sprite, and Pepsi rather than drinking fresh water from an unpolluted stream. Failing that, they will be nearly as content if we must drink water that they have purified and bottled. Also, politicians, most of us, and others are always ready to put "jobs first," whatever grace those words offer the dwellers of a dying planet.

Thus, increasing population and pollution is good for business, and so is the decreasing availability of any free or unprocessed water. It will all be processed and distributed.

With continuing deforestation, and eventual damage to the ocean's plankton, both of which reduce nature's oxygen production, and the dumping of toxic materials into the atmosphere, we will increasingly come to depend on processed air. (And of course, many of these toxic materials, when they return to Earth's surface, go into the water.) The trend toward reliance on processed air has already begun: Only the tip of the iceberg is shown through our use of heaters, air conditioners, air filters, deodorizers, and environmental cleaners, all of which enhance our local (house, car, office) environments and simultaneously (through their manufacture and use) add to the average global destruction. The interdependent downward need/destruction spiral is obvious, and uses will blossom suddenly in the beginning of the twenty-first century. There are already coin automated oxygen dispensing machines in Tokyo.

The tendency for the cycle of environmental damage and destruction is encouraged by pollution related health problems, and by the gradual weakening of the human gene pool. These force an increased reliance on medical assistance, which has an incredibly high toxic waste by-product, and which will be under the production and distribution control system.

It is ironic that all the technological advances that will improve the means of production and distribution, and offer cures for pollution related health problems, themselves add to the bigger Earth pollution problem. Thus, our global civilization hastens to complete a beautifully self-serving spiral of pollution and apparent survival enhancement, while pulling each of us more firmly into its deadly vortex.

Most of us are oblivious to the gradual forging of our own shackles, link by link, which has been taking place over so many hundreds of years, eagerly embracing and habituating to our improved standard of living, as we put ourselves increasingly at the mercy of the controlling powers of production and distribution.

The ruling class need only control one of the fundamental nutriments of life in order to have total control over each of us.

Through production and distribution, they are acquiring control of all the nutriments.

Communication

The evolution of the Internet, videophones, and other communications media offer wonderfully improved global communication. One cannot help but feel awe at the exciting changes that are coming. Just the idea of seeing your friend or family member on the video phone, or being able to simultaneously meet with many of them, brings a feeling of warmth and pleasure that will only be surpassed as these promises are fulfilled in the near future. For the more serious moments, the possibility of entering any database about the globe on a few keystrokes any time of the day, will bring an unsurpassed possibility to amass data for research, amusement, or investment.

These developments depend on the rapidly advancing technologies of communications and computers, between which the boundary is becoming vague. This is similar to the vagueness of the distinction between these two functions in the human brain. In a sense, it would seem that in the symbiosis between Homo sapiens, communications, and computers, a global intelligence is evolving in which each of us plays a part.

Even more wonderful, critical to this evolution and in bootstrap fashion also supportive of it, is the promise of one of the most exciting human developments of all. This improved communication will help tear down millennia old prejudices between peoples, races, cultures, and religions. It will allow understanding of disparate peoples to spread across the globe, possibly bringing to fruition the idealization of the brotherhood/sisterhood of Homo sapiens.

Certainly in some sense advanced communications will bring us closer together. On the other hand, our previously conceived unity may yet elude us even as the image of its fruition forms. The dichotomy arises as we ask who is "us"? Along with the uses of technology, we (individually and as a society) will simultaneously evolve in our manner of living and behaving.

This evolution will be so intimately tied to the exciting technological advances in communication and elsewhere, that its offered blessings may blind us to the dangers.

Sean Connery was considered for the lead role in the Warner Brothers film *Do Not Go Gentle,* in portions of which the director planned to "use high-tech methods to shave 30 years off" his appearance on film. Connery would be looking as slim, vigorous, and youthful as James Bond (007). To "use a new technology to cinematically return" any of us to a more youthful appearance takes time and money. It is a frame-by-frame computer process. (Fleming [not to be confused with Ian Fleming, the author of the James Bond series of books].)

With advances in technology generally, and increases in computing power particularly, this capability will soon advance to where it can occur in real-time. In "real-time" means this: as fast as the original image forms, image processing will accommodate and replace it with the more desirable image. And it will become affordable.

The application of this capability to near future communications is easy to imagine. As Internet (or other) videophone communication becomes standard, real-time beautification software will be offered to mold our image to whatever benefit we wish, as it is broadcast to friends and family over the globe. No matter what your state of youth, beauty, handsomeness, wakefulness, or health, your image will be enhanced.

In time too, intellectual enhancement filters will improve our ideas, words, and witticisms in our own or improved voice before they are broadcast. Similarly, live television shows will use these tools. We will all look good and sound smart, whatever our true selves.

This is just one more example of our increasingly digital existence, and the allure of it. Of course, on the dating scene it may be difficult, embarrassing, or amusing for some of us to ever meet the singles people we have contacted through our beautification filters on the videophone.

Additionally, the ever present vandalizing digital invaders may amuse themselves by inserting hoax viral software that converts our image to make us look like monkeys. Additional viral software could turn our voice into a high pitched squeal or a bunch of chimp chatter.

Politicians too, will be among the first, along with screen stars, to take advantage of real-time enhancement filters. They will increasingly use television and multibroadcast (Internet or other) videophone in preference to live appearances.

In time, because of such technology, in film production the distinction between animation and the use of live actors will disappear. At that time the stellar salaries of stars will decline. Eventually actors will no longer be needed.

In a recent interview, Dave Larson, an executive vice president at an effects company, summed up the current state of the technology:

> He says it's probably feasible today to produce a feature-length film that re-creates a dead movie star, or even creates a human hero who doesn't really exist, "but it would be at an extraordinary price. The only way to do that would be with frame-by-frame animation, and I don't know if any film could recover the kind of money that would take" (Stalter; Johnson).

In analogy with the elimination of actors, sooner or later, either as a hoax or in earnest, someone will promote a presidential candidate who is wholly digital, possibly based on some real person, but where by the time this digital candidate is being promoted, the original person no longer has any part in the campaign. The ruling class itself will present, and possibly get elected, such a totally fictitious digital computer/communications persona. The importance of this is not in whether we will be fooled; rather, the implicit issue is our increasing association and fascination with what is digital, instead of the reality of flesh and blood.

Consider another much more exciting possible impact on politics—the elimination of most politicians. With the average citizen's immediate access to global information and to each other, politicians will become less relevant, at least in casting votes. Their role could be reduced to that of debaters only. An informed populace, after seeing global Internet debate on a given issue, and connected to congressional computers, may bypass politicians to directly digitally cast their own votes, which will be tallied instantaneously, and the result registered faster than in the Senate today.

Of course, power is such a wonderful elixir that politicians will fight tenaciously to keep it. As long as politicians continue to wield their fading measure of power, they will point to every imaginable danger of changing from the old way.

Eventually though, politicians will be pushed out of the loop, and people will enjoy a renewed sense of power and control over their lives in the political arena. Nonetheless, government officials will still exist, as will the ruling class, about whom more is written later.

For all its advantages we will begin to prefer the videophone as opposed to in-person interaction. It will become a bigger part of our lives. And too, out of a need for public safety, and possibly for environmental considerations, it is easy to imagine that in time freedom of assembly will be restricted to digital (virtual) freedom, on the videophone, where any number of people can simultaneously meet together.

There are several negative sides to the evolution of these developments and possibilities. They each concern the extent to which we will lose control of our lives: the issues are privacy, interdiction, and increasing digital existence. To whatever extent we enjoy the benefits of communicating globally from our own unique digital boxes, we also will fall much more under the scrutiny of the ruling class, and will become much more easily isolatable as it suits their whim.

In the present structure of society, it is possible to talk privately. Even if we feel our telephone is bugged, a possibility that the average person at present presumably has no reason to fear, we can meet and talk with one another in person, and be reasonably sure that what we say is private. As our communication is increasingly restricted to the communication media, our privacy will no longer be assured. Our need for privacy has generally been so infrequent, and its guarantee so high, that we have come to assume it.

In the envisioned future, the government, the ruling class, indeed, almost anyone will have access to our lightspeed digital packets traversing the globe with messages of weather, love, hate, poetry, and games. Encryption is the method through which our missives and real-time social intercourse could retain their privacy. This is because an encrypted packet could be nearly as private as an invisible one.

If each of us, each organization, and any corporation could use its own encryption (naturally to be shared with its associates), a similar guarantee to privacy as that in the past would continue. However, we will not be allowed to choose our encryption; rather, it will be dictated to us by the government. (For details about the encryption issue, see the *What About Us?* subsection of *The Digital Battleground* section, some of which is summarized in the following two paragraphs.)

The government, in controlling encryption, will have access to our encryption keys. This will be similar to wiretapping and the judicial procedures that allow or prevent it. Despite assurances of privacy, no one will ever know for sure. And even if the government were to honor its promises of privacy, other sufficiently equipped individuals or organization would not have to do so.

This access, enhanced by advances in AI and computing power, will give the government, and any other sufficiently equipped individual or organization, the freedom to receive, decrypt, monitor, store, retrieve, and review absolutely everything we ever say, send, or receive over the Internet or any other communications media. It may become more than incidental that one such organization will be the communication service provider(s).

Law abiding citizens would hope that there will be no abuse of such power, and that only people who are suspected of nefarious wrongdoing will be digitally scrutinized. On the other hand, suppose the government or some more powerful organization is suspected of wrongdoing, and they do not like our response to them?

More startling than the decline in privacy is the potential for interdiction. The regularity and freedom to communicate (regardless of its privacy) will be assumed. Indeed, with advancing technology, we may reasonably assume that disruptions in service will be infrequent. The service providers will strive to give perfect service, and that should be nearly attainable.

Of great potential impact to our long term welfare and freedom is that we as a people will increasingly depend on these communications services, which will be under the control of these providers. Whether or not we have any present reason to think they would exercise their power, these (provider) controllers can shut service down whenever and for whatever reason they please. If they wish, they may claim "accidental digital failure," "Oops! Sorry!," or whatever, and we will never know the difference.

Thus, in the future, given our increasing dependence on the communication media, we will be more easily isolated at the whim of the the controllers. To whatever extent we enjoy living in a digital box, it can be turned quite suddenly into a digital cage.

Furthermore, given the previously discussed emerging advances in image creation technology, any individual's video communication media presence and that of his correspondents could be synthesized and forged in real-time. Thus, in a national or global crisis any individual or individuals could be isolated, and neither they nor their correspondents will know it until it is too late.

Do any of these conjectures matter? This question may only be answered by a greater understanding of the rest of this book, for it is a conjunction of several advances in technology and consequent changes in society, rather than any one of them alone, which is putting us at risk.

For the moment, it may be sufficient to note that all the foregoing enhancements to communications, regardless of their advantages or risks, constitute one more step in the ruling class's disassociation from us and from acceptance of the reality of our existence. We become more digital, less flesh and blood, as we become more digitally dependent.

Digital Existence

Through the millennia, formulas, algorithms, and digits have become an increasing part of human culture and technology, and have given humans a great deal of control over their environment. This is becoming more true through the digital dependence of production, distribution, and communication.

The flow of data is becoming the lifeblood of our existence, individually and as an aggregate species. While in many ways it provides for a more potent material existence, it also creates new vulnerabilities, and raises the question of new ways in which we may be mortally cut.

Digits are separating us from our past environment, from one another, and from our flesh and blood. Each of us is becoming digitally creatable, and hence replaceable; this is true not only in the video manifestations of our image and voice on a screen, but also in our complete molecular digital specification, and in our complete molecular living presence.

The reliance upon digits for our existence and communication creates a frightening vulnerability, a vulnerability to those who control the flow, storage, and manipulation of data. These include the ruling class, terrorists, and the brilliant insane.

What have we to fear from them?

We are spending more of our time in front of screens, and these are served by digital technology. We sit at communication/computer terminals for entertainment, for employment, and for social activity. Gradually, the beautiful, brilliant, dynamic, screen-saver fish on our screen are becoming more familiar, more meaningful, more real, than any fish in the ocean. And pollution from the technology to create these fish for the abundant and burgeoning human consumer population, will gradually pollute the ocean and sky. This is just one price of our digital existence, but it is not the highest price we will pay.

Just as the digital fish are becoming more real to us, the leadership (controllers) will increasingly see us as nothing more than statistics, already a familiar view to them, but one that is sure to be enhanced by the fact that any of us in our entirety can be digitally stored on, and retrieved from, a disk (as discussed earlier).

Just as we now (by our behavior) deny the needs of other beings outside our group (nonhumans, etc.) to satisfy ourselves, this will be all the more easily done to us by the ruling class as we become less grounded in our flesh and more digitally dependent.

If each of us is stored on disk, and could be retrieved at any time (at the push of a button), the horror of death, or mass murder, is lessened; indeed, to a generation of videophiles it becomes little more than a game.

Through our own will, through wonderful digital enticements and promised technological enhancements to our survival, we are each literally being stuffed into our own individual digital boxes. From this box, we each will have great power to communicate, amuse, and sustain ourselves; likewise, our lives can be completely controlled, and any or all of us could be instantaneously isolated or eliminated by those who are in control. (See also *The Digital Battleground* section.)

More Leisure Time?

In ancient Egypt the pharaohs had a use for their citizens (whether or not they are called slaves) even in the off-season. The annual three months of Nile flooding, when no work was possible in the fields, was an ideal time to work on the pyramids. In the future the average citizen will not be used in creating such edifices, or anything else; there may not even be an Imhotep, the architect. For the future ruling class, the design and creation of the analogues of pyramids will be accomplished by AI, automation, and molenotechnology.

As the bastard son observes in Shakespeare's *Life and Death of King John,* life is not fair, but contrary to another of his claims, the world was not "made to run even upon even ground." There is no divine law of the universe to guarantee that we have sufficient aptitude to keep up with technology, even though its present rate of advance depends largely on our past and present labors. This is a hard, unfair truth, but there can be no veto of its veracity or application.

There is no task presently performed by humans that will not soon be performed better by machines. Examples of this already abound. For example, a cherry pitter was patented in 1863 (Lantz). Of course, at that time someone still fed the cherries to the machine and turned the handle. Humans have subsequently moved further and further from the pitting process, even though they still pick and transport cherries.

Machines will do more than humans can do. Airplanes are an obvious old example. Flight is like a miracle, especially since humans cannot fly as unaided individuals.

Just as machines have surpassed us in pitting cherries and flying, so will computers surpass us in thinking (see the *AI* section). For numerical analysis already one of them working alone can surpass the aggregate coordinated capacity of the human race.

Since the dawn of Homo sapiens civilization, no significant change has occurred in the level of intelligence needed by an individual to fill a role in either the labor or military forces. Throwing spears, roping steers, driving a tractor, sorting mail, and shooting a gun all took about the same level of intelligence, even though in the last century there has been an increase in the amount of training needed. In the next ten years, the increase in intelligence needed to function productively will have no precedent from any period of history.

The advance of technology is occurring at an accelerating rate, and being automated. Automated manufacturing will take place at the atomic level to construct things based on designs of atomic precision. The building blocks will be atoms, even if the product is a jet engine. Furthermore, through AI, automation, and molenotechnology, all invention, design, creation, fabrication, and production is moving toward a sophistication (refinement) in intellectual, creative, productive, and reproductive capabilities surpassing all presently living beings, including Homo sapiens.

Very soon most humans will be far removed from all these processes. No humans will be able to compete with this system; very few will even comprehend it, and fewer still will be able to contribute to its advance, which it will soon do for itself.

The opportunities for employment will decrease as there become fewer activities in which humans can provide a useful contribution or presence.

Occasionally, in response to the above thoughts, concerned people suggest that there will be an increase in the creative arts and personal service employment, such as music and gardening, for example. The trend has been seen, and it may continue for several years, but for the longer term there is an oversight here: It is that the number of people who can afford such services is on the verge of shrinking drastically as the number of available professional service jobs decreases with advancing technology.

Furthermore, in these areas too, AI, computers, and robots will certainly surpass humans very soon, usurping even our cherished creative position, despite its subjective nature. For example, music will eventually fall to the computers' devices and creativity. This trend is already seen in the increasing portion of music which is played today on electronic devices, which can mimic any original instrument and provide a vast variety of new sounds. In the next step, the production of these electronic instruments will be totally automated. Soon too, the computer or an android will play the score. After that the computer will replace the human composers with much more gifted compositions.

The same fate awaits other creative people. With few exceptions, our works will not be highly valued by the ruling class (the only persons who could pay for them). Of course, on a par with how we presently value the works of, for example, Native Americans, there will be a continuing small market for our works.

Consequently, no matter what our individual aspirations, the work week will decrease.

This might be thought of as increased leisure time, if our standard of living were either increasing or at least not decreasing. However, human population and past consumption seem to have pushed us to a turning point. Indications are that Homo sapiens is hitting the limit of Earth's resources, and is still using them faster than they can be replenished or technologically replaced through the use of new materials.

In the era of molenotechnology an abundance of new resources is foreseen. On the other hand, because of initial limited production capability, in the near future there will be a period during which the rate at which these new resources can be used will not grow fast enough to replace the loss of current resources that are now being depleted.

Consequently, our standard of living will decline. Thus, our increased leisure time will be equivalent to more unemployment.

Technology will offer wonderful gifts, but most of us will not be able to afford the significant ones. On the bright side, one might hope that at least our basic needs will be taken care of through automated production and distribution, and indeed this will become possible, whether or not it happens soon enough.

More importantly for our future, regardless of whether or not our standard of living improves, declines, or remains the same, the average citizen will have no function. Too long a look at the enticing leisure implications of this could lead to mortal blindness.

In the past, the masses were needed by the ruling class. The average person was needed for production and to serve as a soldier to protect the ruling class and to protect turf for production. With the sophisticated automation and mechanization of future battlefield encounters, humans will be no more useful than trained laboratory rats would have been to General Schwarzkopf in Desert Storm. As invention, creativity, production, and warfare shift increasingly to the AI, molenotechnology, robotic, and digital arenas, the average person will be incapable of lending a useful presence.

This will be of critical importance in the decisions that lie ahead for the rulers, and it will be referred to later.

Summary of the Technological Environment

If damage to the natural environment is ignored, there are many attractions in the evolving technological environment, though there are risks. Advances in technology have a tendency to excite the imagination of people, just like whatever is "New" does on the shelf at the market. These advances also have the appeal of offering Homo sapiens more control over its environment.

Most popular advances offer improvement in people's immediate lives. On the other hand, such advances offer insane individuals and terrorists greater opportunities to kill people in relatively larger numbers. Also, because of our implied technological dependence, these advances offer the ruling class improved opportunities to control us, since production and distribution will always be choke points that rest in their hands.

In this rapidly evolving new environment, how safe will we be when we are seen as irrelevant to the progress of civilization?

3

NEW ALLIANCES

Ancient Egypt

The nation of Egypt has existed for nearly five millennia. The origin of the kingdom of Egypt dates from the unification of Upper and Lower Egypt around 2700 B.C. Near that time the Step Pyramid at Sakkara was designed, engineered, and built by Imhotep for King Djoser in the Third Dynasty, which ended in 2686 B.C. The great pyramids at Giza were completed by 2500 B.C. Thus, in Egypt in 600 B.C. for example, before the establishemnt of the Roman republic, tourists saw miraculous structures that were already 2000 years old! The stunning mortuary temple of Queen Hatshepsut at Deir el Bahri was constructed around 1500 B.C.; ancient tourists even scrawled hieroglyphic graffiti on its walls.

Ancient Egypt was the archetypal nation—isolated, self-contained, and self-sufficient. It had arable land, sufficient sun for an agricultural economy, the Nile river for drinking water, irrigation, fish, communication, and transport, and mountain and desert barriers to discourage invasion.

Since the time of ancient Egypt (along the Nile River) and Sumer (along the Tigris and Euphrates Rivers), nations have been based on geography. The existence of natural boundaries, which historically restricted the flow of international communication and provided protection, and the presence of fertile lands within those boundaries for self-sufficient food production, have been the major forces in shaping races, religions, traditions, societies, cultures, and civilizations.

The Story of Sanuhi, of unknown ancient Egyptian authorship at about 2000 B.C., illustrates the undying love of the ancient Egyptian for his country (Hatem). After a necessary self-imposed exile, through which Sanuhi often thinks of his country, and after achieving success abroad, Sanuhi still longs for and returns to his homeland at the risk of being put to death. Luckily, on his return he is greeted happily, and lives.

Love of one's homeland, national pride, and patriotism help hold geographically based nations together. They are independent forces, though they have a common origin. They are integrally tied to the safety and validation humans feel in being identified with some surviving group, but just as within urban gangs today, this group success has until now always been based on geographical turf.

Demise of Geographic Nations

The gradual decrease of natural and freely available food, water, and air (nutriments) within arm's-length in one's own geographical region encourages international trade. Water is shipped from the Alps to the U.S. Alaskan glacial ice is shipped to Japan for cooling the drinks of wealthy people. Apples are shipped from New Zealand to the U.S. Regardless of the toxic environmental implications of such profligacy, the nutriment, manufactured goods, and technology trade between nations is enhancing global communications.

Simultaneously, as the need for trade communication is growing, technological advances are occurring at an increasing rate. Some of these make enhanced communications possible. Others offer the promise of future abundance, and generally contribute to the potential for imminent dramatic changes in the structure of human society and civilization around the world.

From the idealistic view of would-be peace-loving peoples, this is a wonderful evolution, one that must lead toward broader acceptance and understanding among all peoples, and perhaps leading even to global unity. We might feel that world peace is at hand.

Our sustenance generally, and particularly the means of producing, processing, and distributing nutriments, are increasingly dependent upon AI, computers, communications, and automation. Imminent developments in nutriment production technology and molenotechnology will reduce the importance of arable land; still resources for processing into artificial food will be needed, but perhaps the kind of land needed will be less specific and more available, and it will be mined and processed like ore.

The Internet and video communications are making it possible for persons to live and work (at a terminal) in a place far removed from their central employment location, where their personal attendance is seldom required, but where the fruits of their labors are synthesized with those of other workers. This separation is analogous to the shift of the average worker from the field to the factory in the 1930s; before that time the majority of us (in the U.S.) lived right where most of what we ate was grown, and we grew it ourselves. We are becoming further removed, by digital technology, from the sources of sustenance. (Though we may become freer to move back to the countryside, it will not be the place in which our particular nutriments originate.)

Thus, with advancing technology, geography and colocation of work place and food source are rapidly fading further as determinants of survival. Likewise, family, race, tradition, and religion are becoming less relevant as group identifiers.

People may rejoice at the imagined eventual demise of racism, for example, but at the same time it could be replaced by something equally pernicious.

Global corporations are emerging. These will control the advancement of technology, and the production and distribution of nutriments, medicine, and communication.

In summary, the means of survival for each individual will become global. Mirroring the global spread of corporations, and to some extent tied to individuals' jobs with these corporations, and also mirroring the global nutriment/trade chain, common interest groups will become global. Nations as geographical entities will cease to exist, or will exist in much reduced importance.

This is analogous to the beginning of the recession of the monarchies and catholicism with Gutenberg's (1390-1468) "perfecting of the technique of printing with movable metal type" (Chambers, *et. al.*) in the 1450s. This advance in the technology of printing made books and ideas more accessible; it marked a significant change in the control of ideas and information. This and other advances led to the industrial revolution, which was accompanied by the rise of entrepreneurs and public politicians. Whether or not beliefs became more correct, their control fell more easily into the hands of this new group of leaders.

Relatively common beliefs, common language, and full bellies have not stopped peoples from slaughtering each other in the past (for example, France at the time of Queen Margot) or the present (Ireland). Regardless of the extent to which control over an individual's or an ethnic, religious, or national group's destiny is the overt cause of such conflicts, survival is the continuing fundamental issue. Survival has always been (1) an issue about which discussion is meaningless without proper context and (2) a relative issue, which together with status, greed, and other factors will continue to drive human turmoil (and destruction of the planet).

Communication and dependence upon computers and technology for survival are allowing and forcing allegiances to shift. What will be the basis of these new allegiances?

Rise of Global Corporations and Interest Groups

Common interests will continue to be based on survival and its enhancements—comfort and pleasure. These historical motivations will define and lead us to the formation of new groups, allegiances, and alliances, because the manner of fulfilling our old needs will be new (as discussed in the previous section).

Many groups will be global and will be in competition with one another. Members of relevant global entities, such as corporations and common interest groups—the new nongeographic nations—will find their self-interest and validation tied to people globally dispersed and united across the internet and other communications media.

The territories and boundaries of these new groups, though less tangible and more digital, will be just as firm as those of past nations. It would be a mistake to construe that all peoples will unite.

Between these new nations the tendency toward war will be no less than in the past, but warfare will shift toward the digital battleground. The digital combatants will be these new nations (corporations and common interest groups), who will find they can do each other mortal bloodless harm through communications media by clever, artificially aided use of digital saboteurs, viruses, bombs, spies, worms, etc. (Doubtless, less technologically advanced nations will continue their usual direct corporal mayhem all the while.) Meanwhile, spies, agents, and saboteurs will abound, illicitly attempting to advance their own group's interests at the sacrifice of the rest, and increasingly relying on digital weapons, molenotechnology weapons, and other technologically advanced nondigital weapons.

Sabotage, theft, forgery, and annihilation of digital property relating to nutriment production and distribution will have as mortal consequences as any past war ever has. These consequences will happen with little more than lifting a finger at the keyboard, and in some cases they will be felt as rapidly as would be a nuclear explosion. There will be no punishment severe enough to deter, nor shield strong enough to stop, all such attempts, and some will be successful.

Furthermore, for those of a more independent, piratical nature, world economic resources will be there for careful digitally based disruption or plucking.

As previously stated, nongeographical common interest will define groups. The powerful enlightened individuals will recognize their individual best advantage and strategy based on their world view. They will express themselves through the power of their corporations or whatever.

Families, Corporations, Gangs, Terrorists, Insanity

Group behavior and our gravitation toward groups will continue to be driven by our desire for survival and its enhancements, and by our ability to comprehend which groups can maximize our benefit. To whatever extent our training and dogma does not guide us to the contrary, our tendency is to choose groups that offer immediate or short-term benefits for ourselves, at the sacrifice of benefits to other beings or long-term benefits to ourselves.

It is a simian instinct to want to belong to a group. This is one of several survival instincts. We are usually born into some groups, such as, family, clan, tribe, church, and state. Throughout our lives most of us tend to remain with, or gravitate toward, groups and organizations that can give us sustenance and protection. This is simultaneously true on various levels of our lives, whether with gangs, clubs, employment, politics, patriotism, or churches. Usually we find simultaneous, sometimes conflicting, interests in and allegiance to several groups.

The radical advance of technology necessarily upsets the balance of groups to which membership offers satisfaction and security in our lives. From a socioeconomic viewpoint, the family structure was understandably very important when the family farm was our whole life—where we were born and where we expected our children to care for us in our old age. The farm was the family business, and it was a lifelong business. Today only a miniscule percentage of the population lives on farms. Following the abandonment of the farm as a way of life, the abandonment of the family has been slow. Of course, the tradition of family loyalty was continued through family dogma, and enhanced by common experiences within the family, regardless of its location or its professional pursuits.

In our label-oriented, communication media, and data driven society, our attention is increasingly focussed on issues having nothing to do with the family. The exception to this is an ironic unnatural admonishment alerting us to the decay of the family and family values; such a warning delivered through the same communication media that are prying us away from the family can have very little impact on our course, which will be determined by the realities of survival. The family was important before, but our very technological and social structure, which we are choosing in our continual demand for more material benefits and their associated necessary advances in technology, are rapidly making the family an anachronism.

Through the millennia of human families in various civilizations, a great deal has been said about children—some of it good and some of it disparaging. Acceptance that youth always go through a difficult period during their adolescence was recently expressed with humorous compassion and love by a parent:

> ...we gave up. And that worked. She's been improving ever since the day we decided to treat her like the family cat—to appreciate her for what she is, feed her, house her, and not expect anything in return (Flaxman).

Writings dating from as far back as 2800 B.C., in the Sumerian culture, are fascinating in how similar they are to some disparaging remarks that we occasionally hear today. For example, the ancient Greek poet Hesiod (circa 700 B.C.) in his *The Works and Days* foresaw a time within the iron age "when the father no longer agrees with the children, nor children with father." He also recognized quite simply that rather than keeping their attention on their work, younger people were always looking for excitement with other young people. The Greek philosopher Plato (circa 428-348 B.C.) in his *The Republic* expressed an opinion that in a democratic society young men would bring insolence, anarchy, and waste into their homes.

Whether the attitude of adults toward youth is always like this, or whether it oscillates through the ages, is not clear, but it is interesting to know these feelings are not new.

A more positive, hopeful, or demanding set of ideas about youth is contained in the following quotes: "Our children are our most important resource," or "The future belongs to our children," or "Our children are the future." These are heartwarming ideas, and they express thoughts whose origin is partly in maternal and paternal instincts; they are also family bonding words, and they are rallying cries for school budget increases. The happy ring of these platitudes may have been reasonable a decade ago, but not today.

Momentum drives these memes forward, but due to global changes and advancing technology they are now founded on four falsehoods.

The first falsehood is based on the mistaken implicit assumption that if children are delightful to know, have, and raise, then having more children is better. In the narrow view of family, without a view to the global situation, this conclusion is still accepted and could prove to be true. On the other hand, the present global population at current levels of consumption and its resulting pollution keeps the planet on a death track. More children, and the increasingly toxic implications of each human's existence in our technological society, guarantees the near-term death of Earth and all its inhabitants who fail to find shelter in artificial environments.

The second falsehood is due to a failure to fully understand the biological implications of health improvements promised by advancing technology. (The extent to which this promise will be fulfilled, which will be discussed later, is irrelevant to the falsehood.) Advances in technology will give Homo sapiens the power to sustain health, prolong life, give immortality to some, and offer rejuvenation to the lucky few. Thus, the biological distinction between youth and age will soon begin to blur.

The third falsehood results from changes in each person's implied or imminent value to society. As previously mentioned, advancing technology will soon render most people useless. In a society in which the necessary rising skill levels are outstripping the average person's ability to learn, a society without gainful or productive activities for the average person, a society in which the employed labor force will necessarily shrink, a society in which soldiering will be largely automated along with most creative and production activity, the importance of youth becomes more negative than positive, and continued birthing may only imply a greater carnage when the bell tolls.

The fourth falsehood is based on the failure to perceive the coming shift in social prejudice about being young or being old. Life time is an asset, a temporal asset completely aside from its corporal implications in terms of earning or productive power. On the other hand, as an element of social placement and status, it is similar to corporal assets or money in the bank. The number of years anticipated until a person's death, the activity level of a person, and the general appearance of health are related to a basic assumption about the gift of being alive and the status associated with having more life to live.

There are few wealthy elderly people who would not trade all their material assets to become young again. Almost every elderly person has some dream age they would like to resume for biological reasons, aside from memories that reside there. With increasing longevity, immortality, and rejuvenation, the temporal assets associated with a given person's age will expand, thus leveling this asset across broader age groups. Ultimately, this discriminator will be eliminated or perhaps even reversed.

The previously quoted positive youth oriented platitudes may never have been much more than verbal conceit regarding Homo sapiens' natural instinct, and politically inspired propensity to increase population for the workforce and soldiering. In addition they served to stroke young people, to help them achieve their identity, and onto whom the mantle would eventually pass. On the other hand, they subvert adult responsibility to move the planet's life course toward a safer path. Thus, they encourage irresponsible procreation. This is all in analogy with, and part of, Homo sapiens' tendency to extol the virtues of aggregate Homo sapiens behavior on the average, regardless of its destructive impact, just as youth gangs do for themselves every day.

These suggestions may seem harsh, or even judgmental; rather, their importance lies in the opportunity offered to review where we are now, regardless of what we have been telling ourselves in the past. We have simply been taught, been loving, and followed what we thought was a good, healthy, and correct path. We did not know where it all would lead. We were just doing what was done and taught before. Why should we have questioned anything? Why should we question now?

Our group oriented gravitation sometimes is very spontaneous or immediate, and other times takes place over a period of years. We have socialization processes that teach us, train us in, loyalty; these bonds serve a useful purpose in stabilizing societies and supporting their survival. The mobility (potential) we experience is limited by the degree to which we perceive truth beyond that which is acceptable within the groups to which each of us belongs at any given time, by our awareness of possibilities beyond those groups, and comprehension of possible benefits weighed against risks. Also, our learned moral precepts, which affect how we choose our groups, can change under financial pressure, for example.

If we find some group to which we wish to belong, the group will assess our potential benefit to it and decide our membership accordingly. Corporations tend to be very choosey about this, since they pay salaries and must make a profit; a church, on the other hand, might accept all comers, if it expects cash donations in return for its low overhead other-worldly offerings. Likewise, political parties tend to be nondiscriminating, since they expect votes and possible donations without having to give anything back; the benefit to the members comes largely from outside the party structure (if the party wins). Once we are granted membership, we are encouraged to espouse the dogma of that group, whether or not all of its dogma had previously been a part of our philosophy (and as part of loyalty, we are taught to stand up for the group, even if we discover it does things that are against previously held moral precepts). Sometimes we adopt the dogma prior to seeking membership, as part of substantiating our candidacy.

As the power of the corporations grows, we will necessarily see our most important interest is in being employed by one of them. On the other hand, the corporations will be formed less of people and more of automation, AI, communications/computers, and data. As the digital and logical assets and operations of corporations becomes an increasing portion of what they are and do, necessarily, the common person will find fewer available places there.

As an increasing portion of us is shoved from all gainful employment, even if initially there is enough bread, we will gradually seek other meaning in our lives. We will still want to belong. We will still want to feel we have control over our lives. And we will want to have things that we don't have, but that the ruling class has.

The ruling class will have no more desire to share with us than we today have a willingness to sell our cars and homes to send food to people starving overseas.

The deciding issue will be raw power. The ruling class will keep what it has the power to take and keep. The rest of us will take or keep whatever we can.

It has always been this way, but the truth of this is made less obvious by the political struggles and smoothing negotiations that in the past arrived at a balance between the various forces in a manner that usually, though not always, prevented bloodshed in the class struggle. These successful negotiations were made possible by the simple fact that the ruling class has always needed most of us, so the issue was really one of price. In each struggle, no matter how heated it temporarily became, both sides at least subconsciously understood that price was the issue, and this understanding guaranteed that an eventual working partnership would be achieved.

The point is that with advancing technology the balance of power is radically shifting in favor of the ruling class, because they will no longer need most of us at any price. When they and we begin to realize that, this continual age-old conflict will climax.

History and current events make clear the response of people in situations of utter hopelessness and deprivation. When people have nothing to lose, previously avowed moral concepts become irrelevant. New ideas arise that offer satisfaction, retribution, destabilization, and a few moments of potent survival, even if the likely result is death.

4

NEW BATTLEGROUNDS

The 1996 agreement among Russia and the world's seven richest nations to end nuclear testing might have been a victory for peace, if it had occurred twenty years ago. Today it is simply a pathetic recognition that among developed nations nuclear weapons are becoming less the weapon of choice in winning the wars that will matter, and an admission that the importance of world political leaders is declining.

There is no value to falling into the political or semantic optimistic trap of those who would pose as doves in the development of "purely" defensive weapons. The technologies of defense in most cases are the same as the technologies of offense. The defensive strategy of deterrence in the nuclear age, for example, at least did not yet lead to nuclear holocaust, and now appears unlikely to do so. However, neither can we ignore that, despite its proclaimed defensive purpose and its success until now, it has also served as an enormous stick in a purely offensive sense. So it is not necessary to focus on whether purely defensive purposes might be a motivator for future weapons development; whether for good or bad, intentional or not, the most purely defense weaponry has an implicit or latent offensive element.

Battles have always proven the power of nations, one over another, and established, continued, or eliminated hegemonies, and so they have historically been used as place markers, and, despite the casualties and destruction, they serve as an important outline for the evolution of world order and the consequent flow of human civilization. This will be no different in the future, but in the past, battles were usually fought over land, in the sky, or at sea, even if the implicit goals were economic.

Future battles will tend to be very different in their composition, combatants, and armament. Acquisition or destruction of data will more frequently be the goal, for which digital and molenotechnology weapons systems will be much more important than guns.

Though the most significant battles of the future will be bloodless, they will yield a higher body count than any previous warfare, including Iwo Jima, Hiroshima, and Nagasaki.

Conventional and Nuclear Weapons

Advancing technology will increase the potency of nuclear and conventional nondigital weapons. In addition to guns, grenades, flame throwers, etc., these include viral, bacteriological, poison gases, and other biological weapons. And we will continue to be at risk to them, especially by less wealthy nations, terrorists, the insane, and other renegade entities (who have less to lose in their use).

In the jungle, some primate species use teeth, fingers, and fists in their tribal wars. Perhaps this was at one time true as well for humans, before clubs, knives, axes, swords, spears, bows and arrows were invented. Likewise, just as these weapons of medieval warfare are now much less frequently used, so too in the future will the use of current conventional weapons decline on the average, though we will continue to be poisoned by their production.

New Nondigital Weapons

Advanced technology will promote the creation of powerful personal and global weapons, imagined previously only by science fiction writers and some not yet imagined. Some of these might be designed by a solitary person sitting at home with a computer, unless there are laws and control mechanisms to inhibit such threatening activities. Others will take a more sophisticated effort, such as that available through the aggregate computer power that will be available on the Internet.

Though most of the examples of nondigital weapons considered below include sophisticated computers to execute their deadly microscopic robotic tasks, their action is directed at physical, rather than digital, interdiction or destruction. Nondigital weapons may have the same targets as digital weapons, but a plastic explosive, for example, would additionally cause physical damage to the equipment (and the operator) of a digital processing system. Also, nondigital weapons may contain digital systems. For example, a guided missile relies heavily on digital systems for its guidance, but in this section such a weapon is called nondigital. (Digital weapons and systems are defined in the next section, *The Digital Battleground.*)

In the following examples attention is focussed on the molenotechnology weapons. The initial weapons developments will include those particularly suited to assassination.

If less drastic criminal decommissioning is required, say in a robbery, a special knockout bullet could be used that is fired into, say, the victim's brain. It kills, but it is composed of molenorobots that use the damaged cerebral materials to repair the brain, and then these molenorobots restart the body, and the victim simply regains consciousness, never feeling, though perhaps realizing, that he has been dead for several minutes. This imagined cerebral repair would not restore the portions of memory or personality that were damaged by the gunshot wound, but it would put the person back on his feet.

Murders executed with assassination weapons will be perfect crimes in so far as the perpetrators will not be traceable. Furthermore, unless the relevant sleuth arrives on the scene during the murder, and witnesses it, the cause of death may not even be attributable to a molenotechnology device.

(1) The Blob

Steve McQueen's big break in feature films came with his starring role in *The Blob,* a story in which an amorphous mobile blob from outer space arrives on Earth at night. At first it is just a bit of goo that inadvertently drips onto a curious man's hand. The man cannot get it off and so he quickly hurries to a local clinic for treatment. During his wait for medical attention, he is completely absorbed by the increasing blob. The blob exits to find its next victim, a mechanic on his back working late under a car. The blob covers his head and face first. It is unsettling and horrifying to a child in the audience, but pretty silly really. The solution, eventually found by McQueen's character, was to freeze the whole thing and ship it to the North Pole, which fits humanity's usual folly of thinking if we can't see something anymore then it won't be a problem. Probably the producers were smart enough to see the possibility of a sequel.

The sequel may come in reality instead of on film. Such a blob creature is entirely feasible with molenotechnology.

(2) The Razor Web

You are walking through your house one day, and before you see anything, you feel you have run into some spider web. As usual, you try to pull it off, and you have some success, but you don't get it all off. In places your skin begins to bleed through thread-thin incisions. The incisions keep getting deeper. By the time arteries are cut, you are bleeding to death.

The web is composed of millions of linked molenorobots that have been programmed to disassemble each of certain molecular bonds in the human body.

A scalpel or razor blade cuts by the incredible shear stress that is applied due to the sharpness of the blade as it is drawn across the flesh molecules. This shear stress pulls the molecules apart as the blade moves.

For the web, each of the millions of molenorobots will attack a different molecule, methodically disassembling it (pulling it apart). The resultant damage to the human body will be about the same as a razor cut, except the cutting will not stop. The robots do not get used up nor do they wear out. They take energy from your blood as they work, and go from one molecule to the next disassembling them.

When these web molenorobots have finished their work they will disassemble themselves into nonmolenotechnology components. There will be no incriminating residual molenotechnology material. The grotesque condition of the corpse will be the only clue to what happened.

The web can be targeted at a specific victim. For this case, before each molenorobot does any damage, it will enter a cell of the person and check that person's DNA. If the DNA is that of the intended victim, death will result.

(3) Poison Droplet

A droplet of a specially designed molenorobots could be put in someone's coffee, or simply dropped onto their skin, which it would easily penetrate. Only a miniscule amount might be sufficient. The molenorobots would enter the blood stream and then gather and attach themselves at some predetermined (programmed) location, say, on an artery wall. There they would methodically disassemble molecules of the artery wall until a hole had formed. Internal hemorrhaging would result in death. As in the case of the web, the molenorobots could be designed to check the victim's DNA before acting; if the DNA was not that of the target victim, they would disassemble themselves.

For those who are lucky enough to be alive in this evolving future world, such a droplet may instead contain molenorobots with computer brains that are programmed to completely disassemble or rearrange some portion of our brains. They would accomplish their mission by carrying out specific molecular assemblies and disassemblies among our neurons. They might make us forget, believe, or do anything. Of course, it might be cheaper just to eliminate the victim through disassembly, make revisions on a data file, and then reassemble the newly programmed person. (See *The Digital Battleground* section for the use of this method for mind control, etc., with molenorobots that we have permitted to entered us for ostensibly benign purposes.)

(4) The Assassin Snake

One of the earliest prototype weapons will be poisonous snakes that have been enhanced with molenotechnology targeting mechanisms. The enhancements will consist of a molenotechnology computer in the brain and the addition of a computer enhanced optical system, which will be implanted in the skull of the snake. These enhancements will effectively boost the snake's intelligence to near that of a human, and it can be programmed to seek and bite any particular person.

The gaboon viper has two-inch fangs. These are suitable for getting poison deep into the victim. Gaboon viper venom, like that of some other poisonous snakes and spiders, works in two ways: it contains a disabling neurotoxin, and it contains another component that begins to turn the victim into digestible goo.

Early prototypes of this weapon will be identifiable as molenotechnology-enhanced only if the snake is captured and dissected. As techniques progress, and the integration of enhancements becomes more sophisticated, an STM (Scanning Tunneling Microscope) will be needed to prove that any part of the snake is not nature's own.

Regardless of detection of nonnatural molenotechnology components, it will be impossible to prove the origin of the molenotechnology material or the assassin snake.

(5) Global Goo

On the scale of global weapons there will be some frightening things too. The rate of replication of organic cells can be duplicated in molenotechnology replication; indeed, it can be greatly surpassed. A specially designed molenotechnology replicator might be silicon based, and if turned loose on a beach could convert the whole beach into replicas of itself overnight.

The right design could convert the oceans to a toxic continuous gelatinous goo within days. Such replicators could be intentionally designed, or they could arise accidentally, just as viruses mutate.

Such molenotechnology creatures would be so alien that nothing could stop them, except a specifically designed molenotechnology defender; they would have no natural enemy.

(6) Brain Invasion

Without your knowledge, a droplet of molenorobots could fall on you and be absorbed through your skin. This liquid team of microscopic intelligent beings would enter your brain. They could take you over. They could literally make you do anything they or their progenitors desire. They could make you lie, cheat, steal, kill, or commit suicide; furthermore, it would feel good to you.

The Digital Battleground

Digital weapons are those weapons whose action and impact is primarily at the digital level. It includes those weapons that can delay, disrupt, steal, falsify, damage, or eliminate stored or dynamic data without apparent physical damage to the digital systems that handle—process, move, or store—that data. Digital systems include computers and any other systems that are commonly referred to as data processing, information processing, and communications systems.

The digital battleground will be horrifying in the eerie silence of the massive destruction that will flood it and spill over into the our lives as innocent bystanders.

Digital Assets

Anthropologists and mathematicians are generally agreed that man's historical use of the decimal (ten-digit) system originated with the fact that humans have ten fingers. If our toes had been better articulated, and we had not usually been standing on our feet during the hunt or transactions in the market place, we likely would have found ourselves using a twenty-digit system. Whether we study time, mathematics, physics, chemistry, or biology, or peer through the other infinite depths of the universe, there is nothing to promote preferential use of the decimal system other than our ten fingers and our historically convenient use of them.

With the invention of the computer, attributed to Charles Babbage (1792-1871), the issue of the best numeric system in that environment presented itself as a choice. Even in the use of fingers for counting, each finger stands for one of two possible things—*zero* or *one*, a *yes* or *no*, an *on* or *off* state. In a fundamental sense, the simplest switches have two states—on and off. Such switches abound in nature at the microscopic level from which they are chosen for modern computers: magnetic polarity is north or south, the electronic gate is either open or closed, an electron is either present or it is not present, the chromophore molecule (of Halobacterium halobium) is either storing a photon of extra energy or it is not.

In nature's organic computers, of which the human brain is but one example, the role of a switch is partially filled by a chemical synapse. Across this junction between neurons, chemicals (neurotransmitters) may excite or inhibit an electrical neural impulse. Each neuron is either excitory or inhibitory; this characteristic is fixed in the endings of the branches of the neuron's single axon, which may carry an electrical impulse away from the neuron and to various synapses of various neurons. At a synapse for a given neuron that might receive and transmit an impulse, many dendrites from the neuron are connected to one or more branches, respectively, of one or more neurons, each of which might be delivering an impulse. The impulse in each axon ending at this single synapse causes movement of excitory or inhibitory neurotransmitters across the synapse. The result of nearly simultaneous impulses arriving at a given synapse is tallied in terms of the sum of excitory and inhibitory neurotransmitters simultaneously put in motion there. If the proper threshold for a net excitory response is achieved, the receiving neuron fires an impulse down its axon; otherwise, there is no firing of an impulse. Thus, as for simpler inorganic switches, the output of the synapse is either a *yes* or a *no*.

The two-state switch is the basis for computer usage of the binary (two-digit) system, which in the classroom and in theoretical papers is usually expressed in terms of 0s and 1s. (For example, what we normally call fourteen and see as 14, which means 1 ten plus 4 ones in the decimal system, is written in the binary system as 1110, which means 1 eight plus 1 four plus 1 two plus 0 ones, which equals 14.) Today, depending on the purpose and application environment, switches of more than two states are also used.

Regardless of the choice of numeric system, anything tangible, or expressed in tangible form, can be completely described in terms of numbers, which are expressed in terms of digits: da Vinci's *Mona Lisa*, Beethoven's *Moonlight Sonata*, formulas, an amulet, Shakespeare's *Hamlet*, a person, etc. All these things, whether we consider them to be merely data, information, or perhaps something spiritually or artistically more precious, can be digitally represented to atomic detail, using molenotechnology if necessary, and consequently can be digitally manipulated, stored, or conveyed.

Music CDs and the videocassettes we use for movies are examples with which most of us are familiar, whether or not we know the microscopic mechanism. In most computers this information is stored in a binary form, in terms of the two digits, 0 and 1, in a string of as many 0s and 1s as is necessary; each place in the string is called a bit. So, *Hamlet* or a human can be reduced to a bunch of bits—a string (a very long string) of 0s and 1s in a particular sequence.

The byte, consisting of 8 bits, is as important today as was the gold doubloon along the Spanish Main in the 1700s. For the convenience of pirates and merchants, doubloons were often divided into eight bits, whence comes the modern usage of bits in computers and two bits in quarters (two eighths of a dollar).

A computer is a combination of switches, switching mechanisms, and transmission paths for conveying bits—the states (0 or 1) of switches—from one place to another within the computer. The bits are held in the switches, as a 0 or 1, or in their dynamic form during transmission along a wire in an electronic computer the bits are composed of voltage pulses (say, zero or five volts, which represent a 0 or a 1, respectively). Data are processed by combining states of various switches to set other switches (to 0s and 1s)—bits are combined to set other bits.

A communication system, in its fundamental parts, is similar to a computer, except that there is more emphasis on transmission of bits, and less emphasis on assessment and combining of bits to set other bits. With the increasing networking of systems, the distinction between computers and communication systems becomes blurred; frequently, several individual computation units may work simultaneously and in conjunction with one another on a computation or processing project.

A data storage system, by comparison, tends to be mostly just switches, for storing bits.

To summarize, all things, information, and data can be represented in terms of digits (and subsequently bits), and there is a distinction between the abstraction of digits and the physical means of handling them. Handling digits (data) includes containing, processing, carrying, sending, receiving, and storing them. In our present technology the physical means of handling may be mechanical, organic, electronic, or photonic.

Babbage's computer was designed to be mechanical. The original ENIAC computer (1946), and subsequent generations of computers, have been electrical. There are now designs for optical (light wave and photonic) computers; some prototypes have been built. In the human brain and throughout the body, the communication system of nerves uses a hybrid electrical/chemical system; the electrical impulses, which are triggered by chemical interactions, travel much more slowly (several hundred feet per second) than electrical impulses along a conductor (at one third to two thirds the speed of light) in a modern computer. In optical computers, depending on the medium of transmission, the transmission speed may approach that of light in vacuum.

Processing speed in computers is affected by several things: how fast (on/off) switches open and close, the speed at which signals travel through the transmission medium, and the length of these transmission paths within the computer. As mentioned in the *Molenotechnology* section, despite the usual slowness of mechanical things in the macroscopic domain, in the microscopic realm there will soon be mechanical computers that are faster than today's electronic supercomputers; this will be due to the incredibly high switching speeds of nanometer-sized switches and the nanometer lengths of the transmission paths within nanometer sized computers. In that same realm, electronic computers will be faster still.

During the Hellenic Age in Ptolemaic Egypt, at its peak the Alexandrian Library contained more than 400,000 papyri of various lengths. That was a major milestone in the evolving importance of information (or intellectual) assets. A significant portion of that collection was burned by Christians in 391 A.D. during the anti-pagan upheavals of that era. Today, all that information could be digitally stored on several CDs.

Data and information represented in digital form on digital systems have become very important in our society. And, as previously stated, they are becoming more so. Inseparable from this increasing preeminence are the associated digital technology and the digital systems themselves, which handle digitally represented data. Soon some part of such digital assets will be the dominant asset of almost every significant organization, corporation, interest group, and government.

In the past, the fortunes of individuals were usually determined over decades, and those of corporations and the fates of nations were determined over decades or centuries, and depended on many decisions and often on physical battles. Now, the fate of a company can be sealed within minutes or hours on the basis of a computer shut down.

Competition and the historical struggle for security, survival, and dominance will continue, but the physiognomy of warfare will be very different in this new battleground—the digital battleground.

Networks

Should a person be allowed to vote if he does not understand networks? Analogously, we may ask if a three-year-old child should be allowed to vote?

Whether or not this analogy is yet appropriate, it soon will be. There are many things a person does not need to know to survive, and many things they may know that cannot affect their chances for survival, and maybe knowing about networks is one of them. On the other hand, from the standpoint of knowing the effect on its own life, a three-year old would not comprehend the difference between a tax measure that doubles the tax on the head of household and one that cuts taxes in half.

To most of us, networks are as mysterious in their workings as a tax bill would be to a baby, yet they will soon have more impact on our lives than any tax bill ever has, and will be just as contrary to the meaning and intent of the Constitution of the United States.

So, perhaps a fairer question would be: "Does your opinion matter?" More on target might be, the question: "Does your opinion matter, even if you are enlightened about relevant issues?"

The world, civilization, and technology are changing rapidly, and mostly in ways that we individually cannot affect, whether or not we can comprehend them. On the other hand, a fundamental precept of democracy is that, as an aggregate, we can control our destiny. The startling and unsettling reality in our evolving world is that our aggregate (majority rules) potency for affecting the course of world evolution is shrinking, probably through processes that cannot be stopped, but possibly that might yet be altered (or retarded), if we so choose. In any case, our potency is shrinking at an accelerating rate just as world human population is still increasing at an accelerating rate.

And networks will be intertwined with these changes.

A network is two or more computers and (network) switches, and connecting communication media to support the transmission of data between the computers and switches.

For clarity, it must be understood that the switches that are used in networks are more complex than the two-state switches that were mentioned in the *Digital Assets* section. These (network) switches are computers, each of whose main task is to guide the flow of data to and from other computers and switches.

Most networks use some combination of wire and glass fiber cable to connect all the elements—computers and switches; however, cables are not needed if all the computers and switches are attached to appropriate receiving and broadcast equipment.

A network usually has more than one user, or has more than one terminal active at a time, and is geographically spread over more space than a desktop.

The Internet is an example of a public network. "Somewhere between 20 and 30 million people around the globe use the Net more or less regularly" (Flynn). "...the number of new loggers-on (is) growing at the rate of about 2,300 percent a year..." (McGrath). On the other hand, some ethernets connect only two terminals and as few users.

A typical modem connected to the Internet allows a rate of data flow of 28.8 thousand bits per second (Kbps), but much higher rates are possible; for example, an ISDN connection to the Internet will allow 128 Kbps. On a typical currently installed ethernet, the data rate is 10 million bits per second (Mbps). Clear telephone conversation takes about 16 Kbps. A videocassette requires a thousand times that rate.

In the evolving public networks, such as synchronized optical network (SONET) with Asynchronous Transfer Mode (ATM—not related to automated teller machines) 155 Mbps will probably be standard. This would allow transmission of all the information in *Encyclopaedia Britannica* in one second. However, just as with Los Angeles freeways, speed also depends on usage—the maximum allowable 55 mph may be achieved occasionally, but high usage implies gridlock. On a network the analogy to gridlock is data collisions and overflow, that is, loss of data. Eventually, networks will be self-monitoring, and will give notice when service is slow, risky, or not available. Also, in analogy with roads, there will be more than one path between two points.

A network may be used simply for communication—transmitting voice, video, documents, data files, or program files between sites—or (at the will of interested users) the network, or some portion of it, may function as a single computer. The exciting promise of networks arises in a seemingly endless list of advantages: video on demand, videophones, global medical consultation at a moment's notice, global 24-hour information resources, global relationships, elimination of politicians and lobbyists (as mentioned in the *Communication* section), global employment opportunities, working at home, global family, global electronic commerce, etc. As the Internet becomes more secure, business applications will blossom (see more on business applications in the *Logical Security* section).

With such a plethora of bright possibilities, one may wonder if there are any risks with networks.

Data will flow in packets of just so many bits at a time. Each such packet of bits will have an address for its destination as well as the sender's address, just as a letter in the U.S. postal system has both addresses. The network will automatically decide the path of each packet to its destination according to traffic loads, etc. Packets from a given sender need not all follow the same physical path over the network. It only matters that at their destination the packets for a given communication are all received, reassembled in their proper order, and that the time delay is not bothersome during, for example, voice communication.

Packets from various network users will mingle through the network. Thus, bits from friend and foe alike will be mixed along the digital pathways, regardless of the confidentiality or sensitivity of the associated data.

The increasing importance of data, and the decreasing possibility of its physical isolation, imply the necessity of potent logical (digital) barriers and boundaries. In addition to a concern for the integrity of these intangible boundaries, will be the equally important question of who controls the network. Unfortunately, with regard to the question of logical security, these questions are like several photons in a light wave—only a small part of its meaning yet inseparable from it. What will be our exposure, what are the vulnerabilities, what are the threats?

Networks will provide the sites of many battles. In the near future, there will regularly be more death and destruction due to digital conflicts on networks than ever occurred in any past battles. On bloodless battlefields more lives will be lost than on all the battlefields of history combined.

Logical Security

The Western horse drawn stagecoaches, carrying strongboxes and rolling under armed guard across the murals on the walls of Wells Fargo Banks, remind us of the importance of guarding hard assets with physical force. Similarly, so do the formidable vaults of such banks. These are the methods of physical security.

In Los Angeles in 1978, Stanley Mark Rifkin, because of his trusted position as a wire room consultant to the Security Pacific National Bank, was able to copy a code from the wall of the bank's wire room. On 25 October he used a pay telephone and the code to illicitly transfer $10.2 million to an account at the Irving Trust Company in New York; the call cost him a dime.

Having made prior arrangements through an intermediary with Russalmaz, a Soviet government diamond brokering agency, in Geneva, Rifkin only had to get the funds transferred to Zurich's Wozchod Handels Bank account at the Irving Trust Company in New York. This transfer established the necessary credit to the Russalmaz account in the Wozchod Handels Bank in Switzerland. In Geneva, Rifkin met with a Russalmaz agent from whom he received the baggage ticket for his pre-arranged $8.145 million bulk buy of graded diamonds. After an airplane flight to Luxembourg, he held the diamonds in his hands. Thus, by mostly electronic means, and without any physical threat or force, he illicitly obtained a pile of an untraceable commodity.

Rifkin might never have been captured, except that his criminal expertise outside the electronic arena was nil: "I didn't have the slightest idea what to do." Almost all the diamonds were recovered by the bank, but apparently Rifkin's purchase was, not surprisingly, over-valued, and the bank lost money in the resale of the diamonds (BloomBecker).

For any business we all know who has to pay for unrecovered

losses. Nonetheless, for competitive reasons, no commercial establishment can survive long in a criminal loss environment. The Security Pacific National Bank has subsequently greatly improved its logical security systems.

The Rifkin affair was just a precursor to the possibilities of electronic and digital crime that lay ahead with telephones, computers, and networks. The maturing capabilities and implied deficiencies in the Internet are no exception, and people have begun to address its vulnerabilities.

> Our security enhancements will make spontaneous electronic commerce possible on the Internet. Electronic commerce can be done today on the Internet, but users must make prior arrangements. We plan to make it possible for buyers and sellers to meet in an open marketplace on public networks and trust each other....
>
> To do this, users must be able to authenticate who they are dealing with. In turn, this makes it possible to authorize access to information, to digitally sign and time-stamp enforceable, auditable documents and to encrypt sensitive data like bid amounts and credit-card information. All of these together will enable institutions to offer commercial services over information highways (Powell, quoting Jay M. Tenenbaum in *Pit Stop on the Infobahn*).

Digital threats will continue and will advance in sophistication in keeping with the technology. As digital assets and activity become increasingly fundamental to our survival and the foundation of our civilization—public, private, government, and commercial—these threats become more important to each of us, and in many cases the immediacy of the consequences is increasing.

What are these threats? What are our vulnerabilities? What are the risks? How can we reduce, block, eliminate, or control them? These are the questions of logical security. How will we defend ourselves?

Though physical theft of hard assets, such as cash and gold, will continue, increasingly the serious criminal will not find much more than adrenalin pumping sport in such pursuits. Even if you get your hands on the asset, it is hard to lug cash and gold around, and in making your getaway you may be forced into a high speed chase by siren screaming police cars, and end up in a gut wrenching crash, from which you will be catapulted into prison. It is great movie material, but not very realistic.

J. Adam in "Data Security" wrote that "industrial espionage... [is] now becoming important enough to cause national concern" (Adam, 19). Bernstein wrote that "North Korea, France, Israel, and Japan are often said to be among the most active commercial snoops" (Bernstein). Though not all such activity is carried out digitally, information is the target, and is thus closely related to our subject.

Examples

To provide a broad context, consider the case of The Chicago River Leak. Pilings were being driven into the Chicago River bed on 13 April 1992. Neither construction workers nor planners were imagining the series of very old delivery tunnels that lay below. When a huge hole was punched in one of the tunnels, many businesses' basements flooded, causing the evacuation of more than 200,000 people. Emergency business sites, including many computer hotsites, had to be set up for more than 30 businesses. This example reminds us that the relevant system usually extends beyond the digital system, it always has a physical component, and it is at risk even when there is no malicious intent (Alexander, from "Computing Infosecurity's Top 10 Events").

Michelangelo Buonarroti (1475-1564) was known mostly as a sculptor and painter. Lesser known facts are that he wrote poetry and was born on 6 March. His birth date is better known now that on that date each year the hard disk of every computer infected with the virus of his name, the "Michelangelo Virus," is overwritten with random data. A special peculiarity of this example is that to a large extent the infection was established by new equipment and disks shipped from Leading Edge Products, Inc., Da Vinci Systems, and Intel Corporation (Alexander).

The November 2, 1988 Internet Worm incident is noteworthy for what it suggests about the mind of perpetrator, Robert Morris, who stated at his trial that he wanted "to see if [he] could write a program that would spread as widely as possible in the Internet." His digital creation wormed its way throughout the Internet so effectively that it shut down many computers worldwide. Later, his father, a computer scientist in the employ of the National Security Agency (NSA) at the time of the incident, described his son as having been bored (Alexander).

At the time of this criminal computer security incident, Robert Morris' father, Robert Morris, Sr., was chief scientist for the National Computer Security Center of the NSA! Thus, there is a grim irony in the notoriety his son achieved with his Internet Worm. BloomBecker even suggests that Morris, Sr. felt pride in his son's achievement; certainly he felt he and his son belonged to an elite group: "I know a few dozen people in the country who could have done it," he said. "I could have done it" (BloomBecker). This is a telling example of the fraternal bond that arises between people whose skills lie among the elements of power of the emerging ruling class, regardless on which side of the law their activities lie in the perception of we average people, whose disadvantage is leveraged in how far behind the laws' evolution we lag.

Definition

The goal of logical security is the protection of digital (or logical) assets—data and data systems (information and information systems).

Still in vernacular terms, but with slightly more detail, the goal of logical security is to prevent any of the following: sabotaging, stealing, damaging, or forging data; illicitly viewing, copying, or deleting data; or impeding the flow of data.

In more technical terms, the goal of logical security for data and data systems, including computers and communications systems, can be summarily expressed as follows: It is the maintenance of availability, authenticity, and confidentiality of data. Availability means the data have been neither illicitly nor inadvertently delayed, diverted, or destroyed. Authenticity means the the data are authentic and have been neither illicitly created nor illicitly altered. Confidentiality means that the data have not been illicitly viewed or copied.

Of course, logical security is not pure; it contains elements of physical security. Obviously, digital systems are physical systems, so physical security must be part of security planning whether the threat is malicious or inadvertent, as the Chicago River Leak attests. Likewise, it is no good for someone to steal your disk.

Just as for physical security, logical security is never perfect; its success is only probabilistic. Generally, expense increases with rising levels of protection. Regardless of the extent to which security mechanisms are implemented, one can never be completely assured that the desired security policy has been implemented fully or correctly, and truly provides the promised level of security. There is always risk.

Threats and Vulnerabilities

Threat is what the offender has the capacity to attempt. Vulnerability is a susceptibility that the defender has; it is an area in which an attack might be successful. It is reasonable to discuss threats and vulnerabilities together since discussion of either one reveals details about the other.

If you use a cordless phone you may already know the significance of electromagnetic radiation (EMR). Anyone who is near enough on another telephone can listen to what you say, just as though they were wired to your telephone.

Similarly, unless your computer has suitable protection all digital activity on your computer is exposed through its EMR, except in this case the eavesdropper must be able to pick out the peculiar electronic fingerprint of your computer from many other computers. Though this has been compared to searching for a needle in a haystack, the determined invader could use automatic search programs on his own computer, and will eventually be able to find and isolate any particular computer's EMR.

In the reverse of what an invader may receive through EMR is what he may send: any electronic computer, from yours to those aboard satellites, could be jammed. In "Cyberterror," Mello wrote,

> An electromagnetic device capable of disrupting computer systems could easily be fashioned from an ordinary C-band radar unit (Mello).

Also as part of the EMR issue, anticipated for the near future, say the year 2001, is remote virus insertion into electronic computers. Neither the target device nor its attachments need to be connected by wire or (glass) fiber cable to the weapon; likewise, no designed radio receiver capability is needed within the target system. Robert Williams writes that almost anything connected to a computer would become vulnerable; as examples, he lists sophisticated radios, switching devices, and sensors. (Williams, R. H.) Insertion could be made at any vulnerable point in a network; subsequently, the virus would infect the network.

This brings us to the issue of viruses, which is the generic word for several types of malicious computer program code: Trojan horse, Virus, Bomb, Worm, Bacterium. These are variously defined by Bowles and Pelaez as follows:

Trojan horse: It may appear innocuous, may offer some useful activity, or may be implanted in a program without modification of the source code (thus making it undetectable by examination of the program code). Once inside the computer (network switch), it may further subvert system security.

Virus: Usually small, typically less than 3 kilobytes in length, and usually attaches itself to another (host) program and executes when the host program is called. It replicates itself, and may also have some further malicious function.

Bomb: Logically explodes upon the occurrence of a trigger condition. Data mayhem results.

Worm: Travels through a network from one system (switch; workstation) to another. Worms can subvert workstation operations, monopolize system resources, and saturate network communications channels.

Bacterium: Tries to replicate, and may attempt to acquire as much CPU time as possible.

(abridged from Bowles and Pelaez)

In subsequent discussion, all these types of malicious computer code will be generically referred to as viruses, except where clarity demands distinction.

Prior to 1996 any virus (malicious code) could only spread through executable (program) files; it spread when an infected file was executed. A new development in 1996 was the first reported case of a virus that could infect and spread through nonexecutable data files.

> The bug can live within a (Microsoft) *Word* document file and spread as it is downloaded or passed about—infecting your computer when you read the text. That puts to rest the old view that only programs, not documents, can spread a virus (Weiner).

"Network intruders and virus writers are growing ever more devious," writes Adam (Adam, 20). Additionally he remarks on new polymorphic viruses that spawn their own mutations and on "automated attacks" now seen on the Internet. Along with technology generally, the technology for writing viruses advances.

Viruses could be used to overwhelm a network with bogus packets. Also, they could be used to reprogram network switches, or overwhelm them with superfluous processing requests, etc.

Not all viruses are inserted with malicious intent. The Drip virus topples characters to the bottom of the screen. Its creator probably thought it was a funny thing to insert in people's computers. The CyberTech virus, described in the article "Stop Me Before I Kill Again" *(Infosecurity News),* infects the system, ridicules the user for having allowed the system to become infected, recommends antivirus protection, deletes all detectable viruses from the system, and then deletes itself. Despite such benign examples, prankster hackers can unwittingly cause serious problems. At the least, the innocent user, on becoming aware of the virus, has no way of knowing whether it is harmless, and is obliged to spend resources to eliminate it. Also noteworthy, a joke virus may inadvertently completely shut down the computer system.

What motivates hackers, whether they are friendly or foul? In a broader life context, many people are driven by the simple desire to overcome some particular challenge, or to gain some form of recognition. Climbers of Mt. Everest are an example. As every classroom teacher knows, some students would rather have negative attention than go unrecognized; the same is true among adults. (As another example, see the later *Encryption* section, where there is discussion of breaking RSA 129, and the offered prize of $100.)

Notoriety is the motivation of some invasive hackers. Others, as Robert Morris, are driven by the challenge, or, as suggested by his father, boredom. Regardless of illicit hackers' lack of compunctions about their trespasses, Melka found something more:

> In fact most of the people (hackers) I talked to actually felt that, when they were older, they would like to get into computer security or work on antivirus research. Almost to a person they loved working with computers and finding out how things worked (Melka).

There is a reasonable suspicion in the security community that we are catching only those illicit hackers who wish to be caught or who are the second and third rate hackers. Given what crimes have come to light so far, one may wonder what the first rate professional criminals are accomplishing, and what they are capable of doing that they have not yet done.

Defense

There are many kinds of security mechanisms that can be employed against hackers, intrusion, invasion, and viruses. Most of these fall under the categories of configuration management (knowing your system), access controls, firewalls, antivirus methods, and encryption. These are discussed briefly below.

(a) Configuration management

One of the most effective defensive tools is good configuration management. It is a necessary part of any security policy: you must know your system if you wish to protect it. It is certain that any digital invader will thoroughly learn your system. If an intruder knows the system better than you do, then whose system is it?

(b) Access controls

The first line of defense in almost any situation is access control. For this, until recent times the world has depended predominantly on physical security, but even so, logical security has long played a role. A combination lock on a strongbox or vault employs logical security: the combination in such instances is a logical barrier.

Vice versa, in our world of the evolving importance of logical security, physical security still plays a role. In the simplest sense of it, we attempt to keep wrongly disposed and irresponsible persons physically away from the computer or network. So, physical security is an important part of access control in most logical security policies. Nonetheless, in many working environments efficiency and economics allow the presence of potentially untrustworthy people in computer areas. Thus, further access controls for digital systems rely on passwords, smart cards, and biometrics. These are the overt front line security mechanisms.

(i) Passwords

Historically armies have used passwords to help members of the same side identify themselves to one another.

Since persons might illicitly enter the wrong area, or even because people who belong in the area in which there is a target computer might not be allowed any use of the system, passwords serve as a deterrent to intrusion. These are also used against potential trespassers who use the telephone in their attempts at entry. Typically there is a password for entry to the system, and each user has his own unique password. A specified number of trials is allowed; if the potential user fails these trials the computer may disallow any further entry attempts and may alert security personnel.

As of 1994, a third of all systems could be broken into by one of the passwords on the published list of the fifty most popular ones. Words like Merlin and Garfield were on the list. The average password was being changed only every sixty days. In the Roman army they changed their passwords every day.

Many ingenious methods have been used to break password controls, but when used smartly they are still an important part of access control.

(ii) Smart cards

Smart cards tend to look like credit cards. They contain a computer chip that interacts with the desired computer system to establish your identity before the computer will allow you to use it. A password is usually necessary in conjunction with a smart card. Furthermore, the interaction between the card and computer is different each time, so illicit EMR eavesdropping would not help a potential intruder learn how to duplicate your card.

(iii) Biometrics

Biometrics use hardware and software to assess biological characteristics to identify and authenticate a potential system user.

Retinal scanning is one example. Each human has a unique retinal pattern, just like a finger print. The retinal scan pattern is compared with those stored on the system to determine if the potential user is qualified for system access.

Another example is finger lengths measurements. The potential user offers his hand for automated measurement. The combination of five finger measurements distinguishes each individual.

An example of a subtle biometrics method is typing pattern recognition. When the potential user types in a password, name, etc., the recognition program begins assessing the dynamics of the keystrokes. These dynamics are compared with those on file within the computer. Access is accordingly allowed or denied as appropriate.

(c) Firewalls

The word firewall originated with a wall that could impede the spread of flames in a burning building. Its meaning now encompasses any physical or logical barrier that can limit or impede the spread of a problem.

Within computers and networks examples of firewalls are any combination of hardware and software that prevents or limits the spread of a virus, confines a trusted user's activities to specified logical regions, or prevents specified kinds of data from moving between specified logical or physical portions of the system.

Firewalls are used with computers, file servers, memories, and networks.

As a practical example, suppose a terrorist wished to overload a network with bogus data packets. The terrorist might copy many real packets that were already in motion on the network; there would be no disturbance to those packets. Then, at a later time the terrorist would replay many iterations of these valid packets on the network. A firewall, might, for example, check a timestamp, which would show that the packets were no longer relevant, and then the firewall would dump them.

Viruses can subvert firewalls. So again, nothing is absolutely secure.

(d) Antivirus methods

There are many ways to resist viruses. Rather than diverge from the main course of this introduction to logical security, we proceed to one of the most important antivirus tools, encryption, which has many other logical security functions as well.

[In writing the *Logical Security* subsection, the author gratefully acknowledges the use of a 22 September 1994 unpublished paper, "The Digital Battleground in a Candidate B-ISDN Communications Segment of a Future (2001) AFSCN," now owned by Lockheed Martin Corporation and which he wrote while he was an employee of Unisys Corporation (Network Integration Contract).]

Encryption

The Spartans of ancient Greece employed a staff and a leather thong for encryption. They would wrap the leather thong evenly and tightly about a staff. Then on the leather they would write their message lengthwise along the staff. Once the message was completed, the thong would be removed from the staff. The staff and the thong would be sent to the recipient along with appropriate directions for rewrapping the thong so that the message could be read.

Historically encryption has often been used for confidentiality. Certainly, too, where authenticity was more important than privacy, a message would be encrypted, or a secret code word or symbol might have been attached.

> For the first time in recorded history, the interest of the commercial sector in secret codes begins to rival that of government, and dwarfs that of both church and lovers (Murray).

In our digital civilization, as the Internet and SONET/ATM increasingly become the fundamental medium of our social, entertainment, political, economic, and commercial intercourse, and the competitive speed of our commerce quickens, we will necessarily rely on encryption for privacy, authentication of documents, and identification of ourselves and the parties with whom we communicate.

Generally, on digital systems, encryption can provide protection from a variety of threats. Encryption is used for digital signatures, to inhibit illicit entry, to increase the authenticity and confidentiality of data, to identify intrusions and illicit data alterations, and to resist viruses. For example, inside a computer encryption can be used to thwart invaders and viruses. If all of a computer's files are encrypted, then neither an invader nor a virus can function there without the encryption being broken first.

Just as fire has warmed us and the wheel has carried us, encryption will become fundamental to the struggle for stability of our digital society. It will become pervasive, and then necessary to, though not sufficient for, our survival.

All encryption methods consist of two components: an algorithm and a key. The algorithm tells how to use the key and the cleartext (of message or data) to create encrypted data (ciphertext). Decryption reverses the process.

Methods vary greatly, as does their publicity. For some of them, both components are kept secret. For others, the algorithm is public, where reassurance of its strength comes from peer review, and the key is kept secret. Many people in the encryption community argue that, for any worthwhile method, the detriment of publicizing the algorithm is always outweighed by the benefits of peer review, through which an algorithm's weaknesses can be exposed and corrected.

> Someone who has the ciphertext and the algorithm but does not know the key cannot read the message. In any good encryption system, all security is based on the secrecy of the key. There is no security in (the privacy of) the algorithm. In fact the algorithm should be public, not secret, so that other people can examine it and verify its security. If a company refuses to reveal the workings of its encryption algorithm, claiming that it is proprietary, you should immediately be suspicious of their products... (Schneier).

Also, it is only through an algorithm's public disclosure, at least to its users, that there can be reassurance that the encryption provider has not installed a backdoor through which to do illicit decryption to the detriment of its clients.

For public-key cryptography (PKC) there are two keys, of which one is public and one is private; the algorithm might or might not be public. Rivest, Shamir, and Adleman in 1978 created the RSA, which is an example of a PKC with a public algorithm. The first PKC was invented by Whitfield Diffie and Martin Hellman in 1976 at Stanford University. Part of the beauty and importance of a PKC is that there need be no prior private communication between two parties to have encrypted communication between them!

Your digital signature, which depends on PKCs, is as simple as your name. It is reasonable that you will be legally bound by your digital signature, carried over the communication medium to someone you have never met in person.

If your signature is digital and on the Internet, can someone simply copy it? No. Through encryption it may be more difficult to forge than your written signature on a piece of paper. That is one of the beauties of an encrypted signature. If, for example, you wish to sign a letter to the editor of a magazine, your digital signature and the letter are encrypted together through a PKC. Both would be available for anyone to decrypt with your public key, but no one could surreptitiously alter the letter or forge your signature. Unless an adversary learns your secret key, forgery and alteration are impossible. Similarly, from the standpoint of business and commerce, without fear of compromise you will apply your digital signature to any confidential document, say a contract: after you encrypt the whole thing with the intended recipient's public key, only the intended recipient will be able to read the document and verify that is is from you. These are some of the miracles of PKCs!

Of course, there is no encryption method that is guaranteed to be unbreakable. On the other hand, in the public eye the possibility for the existence of unbreakable encryption comes from the fact that ancient Egyptian hieroglyphics (not meant as a secret code, but the point is the same) were not translated until after a French soldier's 1799 discovery of the Rosetta stone, which contained hieroglyphic text and parallel translatable Greek text that served as a key. Even so, despite great effort, hieroglyphics were still not broken until Champollion did so in 1822.

Gilbert Vernam's encryption method (invented in 1918) has been proven to be breakable only if the decryptor can steal the key (a random number sequence), or if the decryptor knows that the same key was used on each of two intercepted messages. This method was used on messages between Fidel Castro and Che Guevara; it was also used at one time on the hotline between Washington D.C. and Moscow. Vernam's algorithm is brilliant and trivial; security depends completely on the secrecy of the key. Thus, this method would be very useful for some encryption purposes if keys could be easily exchanged with absolute confidentiality. Whether or not the Einstein-Podolsky-Rosen effect (a quantum mechanical effect) will easily allow secret key transmission without surreptitious key compromise remains to be seen.

As far as the encryption community is publicly aware, every currently used encryption method is breakable. On the other hand, encryption is still useful because the encryption breaking (illicit decryption) is not usually practical. Most illicit decryption focuses on finding the two prime number factors of an enormous numerical key.

The issue is always one of technology, time, and money. In a cost effective strategy, the encryptor must gamble that the chosen level of encryption will protect something long enough (minutes or months) to maintain the time dependent value of, for example, the data's confidentiality.

All encryption methods cost in several ways. There may be licensing fees in their use, and they take up processing time on the computer. They delay communications: encryption methods take time to do their work, and some of them send out more bits than were in the original message. Furthermore, there is overhead in key and other system management features associated with encryption and (legitimate) decryption. Also, there is always the danger that a data file will become inaccessible through faulty encryption. It is approximately true that the larger the key is, the more expensive the method is to use, and the more difficult it is to break.

In 1994 a global, Internet-distributed processing team of 600 people, using 1,600 computers, broke RSA 129 (O'Connor). RSA 129 is an encryption method that uses a 129-decimal digit key. In addition to the size of the project, these various computers had worked on the problem on a spare time basis since the summer of 1993; the key was broken into its two prime factors in April 1994 (Powell, "Today's Math Quiz"). This decryption success does not ruin RSA generally, or even RSA 129 (with a different key); rather, it suggests use of larger keys for those secret data that might merit such an expensive repeat proven decryption performance.

This example is significant for several reasons. First, it is an excellent example of using an interacting portion (1,600 computers, etc.) of a network as a singe computer. Second, the size of the prize, $100, was miniscule compared to the significance of the feat. This is a statement about the personalities who inhabit the encryption community. Third, in 1977 RSA's inventors (Rivest, Shamir, and Adleman) predicted that it would be 50 years before RSA 129 was broken. It took only 17 years. This is an example of how technology has advanced faster than anticipated.

In a technology in which the price (price/performance ratio) of computing power is falling, the encryptor is able to apply increasingly sophisticated encryption. The illicit decryptor receives a similar, but not equivalent, benefit. It is a happy note that present trends suggest the encryptor is favored more in the balance.

[In writing the *Encryption* subsection, the author gratefully acknowledges the use of a 22 September 1994 unpublished paper, "The Digital Battleground in a Candidate B-ISDN Communications Segment of a Future (2001) AFSCN," now owned by Lockheed Martin Corporation and which he wrote while he was an employee of Unisys Corporation.]

What about Us?

In this increasingly risky and turbulent digital domain, what will be the place of average people, individually and as an aggregate population? Will we parade our expanding virtual magnificence across brilliant, scintillating, digital landscapes, move only with halting steps through a bewildering digital labyrinth, or find no place to hide from the wild beasts of the savage digital jungle?

To receive harmless entertainment value from this new world will be simple and nonthreatening. What more will there be? What of our communications, attempts at information synthesis, development activities, dependence for nutriments, and freedom? In availing ourselves of the benefits of this new world, we know there will be significant digital exposure. What of our defense?

As implied in the *Encryption* section, encryption is the most important security mechanism within our evolving digital environment. We will all need it; we will all use it. It is also thought that with advancing computer power and its decreasing price/performance ratio, the encryptor is favored over the decryptor, that is, encryption is favored over illicit decryption, as was stated in the previous section. Theoretically this should be excellent news.

Au contraire!

This claimed improving advantage of the encryptor over the decryptor would be very reassuring were it not for several simple facts: (1) it is most applicable between adversaries of comparable wealth and digital computing power; (2) in the future, disparity between (a) computing power and AI that will be available to large organizations (such as corporations, cabals, and governments) and (b) that which will be available to the average person at any price will be legally and covertly increased; (3) the government would like to impose encryption restrictions.

These issues all strike a blow at our privacy and expose us to other previously unforeseen threats that arise due to simultaneous technological advances.

In 1993 the U.S. government proposed a severe legal restriction on all telephone encryption.

> "Recent years have seen a succession of technological developments that diminish the privacy available to the individual," (Martin) Diffie stated last month in testimony before the House science subcommittee. "Crytography is perhaps alone in its promise to give us more privacy rather than less. But here we are told that we should forgo this technical benefit...the government will retain the power to intercept our ever more valuable and intimate communications" (Peterson quoting Diffie).

A public debate raged through 1993 and 1994 about the potential for invasion of privacy of individuals, interest groups to which we may belong, and corporations. Originally there was some concern that such an imposition would be made into law. Possibly that pressure will return in the near future. Regardless of the failure of the government's original proposal, it showed an intent that doubtless continues on a hidden agenda.

Enactment of laws are not the only ways governments execute their agendas. Encryption restriction might evolve based on the government's tremendous commercial clout in the market place. By simply determining to use its own National Security Agency (NSA) Clipper chip encryption standard in its commercial activities, it encourages the acceptance of its standard and our exposure to it.

Of course (as implied in the *Encryption* section) the case in favor of some method of telephone encryption is very strong since most of us would agree with the White House that such "technology would prevent drug dealers, industrial spies and others from intercepting communications...." Furthermore, the restriction to use only the government's Clipper chip in telephones would have the benefit of "enabling law enforcers to eavesdrop on suspected wrongdoers" (Alexander, "Controversy Clouds New Clipper Chip"). Unfortunately, the restriction would make the rest of us similarly vulnerable to the government and a variety of nefarious digital adversaries.

Regarding the encompassing issue of government tapping of digital communications, Dorothy Denning, professor and computer science chairwoman at Georgetown University, offers the following opinion:

> ...wiretapping is an important tool for law enforcement. It's the only tool that can be used to deal with certain criminal cases that can't be solved by other means, for example, in major cases involving narcotics, organized crime and terrorist activities.... If we lose that capability for electronic surveillance, we're potentially opening ourselves up to a situation where we won't be able to deal effectively with certain serious crimes (Garon, quoting Denning).
>
> She says that "technology...could shift the balance away from effective law enforcement and intelligence gathering toward absolute individual privacy and corporate security. The consequences...would pose a serious threat to society" (quoted by Wayner).

The government would like to disallow the use of all voice encryption except for an encryption method invented by and specified by itself. Particularly, we would be restricted to an encryption method (Skipjack) invented by the NSA. Their access to any of our particular keys would hypothetically be limited by judicial processes (that would retrieve them from escrow protection only under court approval). Thus, in analogy to wiretapping of a telephone line, our privacy would be nominally protected as it is presently.

It is interesting and suspicious that the government's initial intent was to keep the algorithm proprietary. However, bowing to public pressure, NSA offered to have some mutually trusted public experts examine the Skipjack algorithm, to presumably report back to the public on the security of the algorithm and the absence of compromising backdoors. Of course, since the Clipper chip itself, with the Skipjack algorithm already programmed in, comes as a unit in the telephone, it would take the eventual reverse engineering of tamper resistant computer chips to discover, perhaps some years from now, what really is on them, that is, it is not sufficient to examine the algorithm on paper.

This encryption restriction issue is not limited to telephone conversations. Capstone is the government's Internet analogue to Clipper, and it also uses the Skipjack algorithm. Network data rates will increase beyond the present 12 Mbps limitations of Capstone; for these the government will be ready with the Fastlane encryption restriction for SONET/ATM.

The government's interest in establishing these limitations, and consequently having access to all our communications, is promoted in the interest of law enforcement agencies who need access to the communications of drug cartels, organized crime, terrorists, the insane, cabals, etc. Such law enforcement might justify the need for the encryption restriction. Indeed, we will probably soon accept this need as being higher than our need for privacy. It will only take a sufficiently abominable digital terrorist attack to prove the government's point. They will be ready to jump in with their encryption solution and restriction at the appropriate moment, and it will become law.

Regardless of some benefits to law-abiding citizens, our exposure in this circumstance should not be underestimated. First, whatever encryption method is used, data packet addresses will be cleartext, so the government will likely have the means to intercept, record, and store any, and possibly all, communications, regardless of their being encrypted. Presumably there will be no nominal need for legal injunctions against this. That would make any of our (encrypted) conversations available for immediate or later decryption and perusal. As mentioned previously, in the *AI* and *Molenotechnology* sections and immediately above, enormous computer power, AI, and other digital tools for such tasks will become available in the future. Second, since the encryption method was invented by and known to the government (NSA), the government can break the encryption relatively easily. NSA itself has stated that knowledge of the algorithm would allow cracking any encryption that uses it (Alexander, in "Controversy Clouds New Clipper Chip"). Also, direct legal access to the keys becomes much less relevant, considering the ease with which the government could crack the keys. Furthermore, the NSA may have installed a secret backdoor, which they will of course publicly deny. The list of possible faults goes on (see Alexander, in "Controversy Clouds New Clipper Chip"). Whether or not such decryption would be legal, past abuses suggest it would happen. Third, with only a limited encryption method available, we would also be similarly vulnerable to any other sufficiently clever and powerful interested parties. It must be remembered that in the past, physical movement of persons and material were usually required to establish such analogous eavesdropping on telephone lines. Those physical limitations afforded us a measure of protection. In the future, all such intrusions can take place from a keyboard anywhere in the world.

This will be an incredible increase in (1) government (and other ruling class) control over us and (2) our vulnerability to other aggressive hostile entities.

Regardless of the obvious legitimate motive of the government for wishing for encryption restriction, there are additional motives.

First, the government is an organism itself, and like all organisms it wishes to survive—that is fundamental to its behavior. In our rapidly increasing digital technology, government sees its role potentially diminishing, and it wishes to assert itself, whether or not it has any value to add.

Second, imagining we still existed in the past nondigital world, the government would feel most comfortable if we would magically earn income, magically buy things, spend all our time in our houses, and pay taxes all without a fuss; of course, we would come out to fight wars occasionally. In the digital world, we will spend most of our time in our digital cells, and we will be interacting across the globe, but were we to use encryption schemes of untrammeled power, we would freely spread our wings and interact with other individuals in a global manner that would tend to neutralize past patriotic ideals and that would supercede the typical bailiwick of past government interests. From the government viewpoint, this would be analogous to us traveling about invisibly and silently; the government would only be cognizant that a lot was happening that it knew nothing about. It is the kind of thing that would make government people feel impotent.

Third, in the context of the shifting elements of power within our changing technological environment, the present form of government has enough collective intelligence to see or sense, subconsciously or otherwise, that its role is in danger of being largely usurped by entities of incomprehensible complexity, which wear masks of encryption, and whose development is driven and understood by people (the controllers) who are gradually moving beyond current government control.

Thus, it is clear why the government wants encryption restriction: in addition to a legitimate concern, it feels threatened. Encryption restriction answers all four concerns, whichever is the more pressing.

Freedom is always contextual, always relative. As our digital existence expands through our necessary reliance on it for our sustenance and communication (and whatever employment there might be), it becomes a great deal of who we are. Certainly, it will be expected that we enjoy its fruits to the maximum allowable extent. It will be the main socially acceptable way in which to expand ourselves as individuals; it will be the way most touted by our peer group. But if we are not allowed maximum protection through encryption of our choice (with maximum guarantees), then each person is more exposed to all potential aggressors, and due to that threat our freedom shrinks, while that of the ruling class expands.

Someone might argue that the ruling class itself (whoever emerges as belonging to it) will be subject to the same encryption restriction and scrutiny as the rest of us, and that consequently they will be just as bound as we. Whether or not the ruling class will be subject to the same restriction, it is patently false that they will be just as bound. If they are similarly limited in encryption, which is unlikely, they will be exposed among themselves, and, to be sure, among other illicit decryption aggressors; however, their possible exposure among themselves does not expose them to us, even though we will continue to be exposed to them!

Their privacy will be largely protected from us, regardless of how they may abuse their ability to snoop among themselves. The exposure only goes one way! Thus, in a relative sense the strength and power of the ruling class will grow. This may seem like a subtle issue now, but that present sense of it is due to our not yet having fully emerged into the digital world, even though it is rapidly engulfing us. In time it will become an enormous issue. And whatever control is established over us now will be inherited by the evolving future ruling class.

Still considering encryption restrictions in communication media, beyond the above dangers to us is another ominous one whose force is multiplied by several factors, including our increasing reliance on the communications media. In these media, we will protect the authenticity of our epistles, voice communications, and videophone interactions by our (encryption dependent) digital signatures, whether or not we encrypt the communications themselves. Furthermore, as mentioned in the *Commu-*

nications section, in the near future computer processing will be so rapid and AI so advanced that it will be possible to create any given person's digital dynamic persona in real-time. Thus, with our susceptibility to decryption, any hostile power, person, organization, cabal, or other entity can take over anyone's digital signature to digitally replace any of us in the communications media with a real-time dynamic vocal/visual forgery, which they create and control and to which they will apply the appropriate forged digital signature. In the media, no one will be able to distinguish this forged digital persona from ourself.

Thus, encryption restriction in the communication media will imply startling vulnerabilities. Where else might an encryption restriction be applied and create danger for us?

Until now, as far as we are aware, the most significant viruses, for example, are those that have emanated from the minds of individual hackers, but as previously noted, it is the lesser ones who have been caught. Such people, and those who are more skilled, continue their virus development activities, and more sophisticated covert development of viruses and other invasive tools is underway within corporations and any large organizations that must compete and value their survival.

In the future, individual citizens will be economically and possibly legally disallowed from any potentially significant development activity. Whether to breathe a sigh of relief or not may depend on current illicit hacker's little understood violent needs and other needs for expressions of potency; the present successes of illicit hacking might be manifestations of motives and psychological needs that should not be too quickly ignored. We must consider the activities to which hackers will turn their attention if they are frustrated in their present mysteries.

It is of overwhelming importance that, to whatever extent we have the technological capacity to develop or use (commercially developed) digital means to protect ourselves, there may be inhibiting legal restrictions on the digital activities of individuals. The groundwork, willingness, and possible legitimate necessity of government to police this arena has already been indicated above in the past proposals for encryption restrictions on the global communications system(s). And these restrictions will be imposed or accepted, as already stated, in the interest of mutual public safety.

An equally important issue, a further binding one as far as our freedom and individual safety are concerned, is that this control—legally imposed encryption restriction—will be expanded to include the whole computer system in our homes. We will be limited within our own systems to using only the allowed encryption protection. (The *Encryption* section mentioned the use of encryption for protecting a computer from illicit access and viruses.)

From the standpoint of public safety, this will make great sense. Ruling class watchdogs must have access to everything on anyone's computer system to reduce the likelihood of the wrong people creating vandalistic or terroristic malicious viruses, and other digital aggression mechanisms. Likewise, they will need to inspect for potentially dangerous molenotechnology designs. On the other hand, this leaves us naked.

Despite any legal guarantees to the contrary, we will never know when these people, software invaders, or watchdogs have been in our system on the pretext of searching for dangerous viral programs and other invasive digital tools. Equally significant, whether or not the government has need of such control over our lives or can be trusted with it, we will be similarly exposed to other malicious entities and organizations who wish to take advantage of us, and who will be able to do so more easily because of the enforced encryption restriction.

Consider for example, what may appear to be a relatively benign invasion, such as an uninvited market survey. Through the Internet a game corporation sends a virus into your home computer, after having broken your limited encryption. For a month it registers all the movies you watch, it checks all the games on your system, monitors which ones you play, and with what frequency you play them. It invades your biometric software and studies your keyboard (and joystick) activity and intensity. All this information the virus reports back to the corporation, and then it erases itself. Subsequently, the corporation assesses this information, combining it with that from other home systems, and designs a game that has a higher likelihood of being purchased.

This might be seen as a happy evolution toward a more slick society, one in which your needs are satisfied without you even having to know them yourself. But maybe you do not wish to be known so intimately. That is the significance of privacy and freedom—your choice. If your choice is being surreptitiously subverted, you have none. We will not have a choice; the extent to which the ruling class (through governments, corporations, or cabals) will prey upon us will increase.

The history of nations and personal relationships has repeatedly shown the enormous advantage that falls to those who know about us, and about whom we know little. In totalitarian states, the exploitation of this advantage is the motive for using censorship, propaganda, and secret police. The goals of these will be more easily fulfilled by the new mechanisms of the digital world, especially as our digital pigeon encryption wings are clipped.

Unfortunately, our plight is worse than implied above.

In addition to our loss of privacy, the constriction of our spirits due to this awesome increase in exposure, and the relative increase in our enslavement due to the increased control over us by the ruling class of this digital world, we will be at risk in many other ways—as intended targets and as innocent bystanders.

It is stunning to think that the poisoning of Earth, and with it the inadvertent body count (the consequent human deaths and species extinction), will necessarily increase in response to the survival and dominance conflicts that are evolving now on a highly sophisticated and technologically advanced plane to which few of us would ever be allowed access, even if we were to survive. Certainly the average person will have nothing to gain. And for those of us who do survive, we will suffer the long-term effects of a declining natural environment, due to these conflicts whose goals will be as irrelevant to our well-being, or as destructive to it as our past conflicts have been to deer in the forest. Equally unsettling is the potential that we may directly die in the digital cross fire.

As one example of the kind of attacks that may be coming, consider a planned attack on the metals market. The carefully designed virus enters the global market system. The human culprits will have digital identities, and so forth, all of which will be erased after the completion of market manipulation, sales are done, profits made, and money electronically entered into secret bank accounts. The culprits are never identified, albeit any parties who profitted noticeably will be suspect.

As another example, terrorists can quickly get the attention they seek by a digital attack on the nutriment production and distribution system. Terrorist attacks usually aim at affecting world opinion in a balance between fear and angry backlash. As the average person becomes less relevant in the scheme of the world, and our lives consequently become less important to the ruling class, there will be a tendency for more people to join such organizations. At the same time a greater number of us will be allowed to suffer attacks by them, before terrorists must beware the angry backlash of the ruling class.

Is it all so scary? Is there no hope? Will attacks be so easily carried out?

In the history of armament, offensive and defensive technology developments have tended to balance one another only at the level of combat between individual soldiers or otherwise localized target-specific combat entities, such as an aircraft carrier under attack by warplanes. For example, an air-to-surface missile aimed at a ship is susceptible to digital interference or counterattack by an anti-missile missile. On the other hand, a bomb, or nuclear explosion, tends to overwhelm everything for miles around, and have insidious long-term effects on both sides of the conflict.

Perhaps the technological balance between digital offenders and defenders will continue as it is, with each side now and then making a breakthrough to invade, foil, evade, or block the other.

Then, the opposing side will catch up, etc. In this manner, corporations and other groups will constantly struggle for superiority and security.

There are several ways in which we will be very much at risk in these optimistically balanced battles between corporations, cabals, terrorists, and the insane.

These digital combat technologies, like technology in general, will become so sophisticated that their development will eventually be limited to wealthy organizations. AI, etc., will be required and humans will gradually fall further out of the development cycle. Likewise, due to cost, and in some cases due to legal restrictions, their use will be limited to government (or the relevant ruling class), or to organizations that are wealthy enough to surreptitiously circumvent legal restrictions on various offensive and defensive capabilities. Nonetheless, the corporate entities, and other sufficiently wealthy organizations, will continue the digital struggle, and the constant research for more effective digital weapons.

The speed at which these digital technologies in general, and specifically those of warfare, are evolving is increasing, and the rate of increase will grow further through AI and the increased speed of molenocomputers. This increases the chance, though not necessarily the duration, of some imbalance occurring.

Suppose that during such an imbalance of forces, the superior force attacks the digital nutriment processing and distribution system. Through a viral attack on digital systems, initiated at a keyboard, delivery of air, water, or food could be cut off by any one of several entities. Possibly the culprit could not even be identified, but any of the possible perpetrators—corporations, cabals, terrorists, insane, religious extremists (for example, the Japanese group Aum Shinrikyo)—could blame any of the others. We will be highly exposed due to our reliance on digital systems, our demand for nutriments, etc., and consequent demand for supporting resources.

It is just as easy to imagine that the digital portion of our nutriment system is inadvertently deleted in a digital attack in which it is not the target. This is analogous to the physical situation in which a hospital stands next to a munitions factory during a bombing raid. In this way or another, in some digital conflict, we may become innocent bystanders who inadvertently die. This certainly becomes more likely as our reliance on digital means of nutriment production and distribution grows, falls more under the control of digitally conflicting entities, and the nutriment stock is reduced and monitored at all locations, including our own homes. Toplining our vulnerability will be our uselessness to the ruling class or any other warring entities.

We must remember that until now our relationship to the ruling class has been symbiotic. Once we no longer have a productive role in civilization, we cannot expect simple humanitarian concerns will save us. Among any elements of the ruling class and any other warring entities, in their struggles for supremacy, the motivation for assuring our safety will decline.

There are further examples of digital dangers. Some of the most startling digital weapons will be those that depend upon unseen microscopic molenorobots for their delivery.

For example, to achieve computer privacy you may imagine disconnecting from the world web, or just having a computer that is never connected to it. You may feel then that you will achieve your desired privacy. This is false. Microscopic molenorobots carrying a virus can climb right over your keyboard and into your machine. Once inside, they can load the virus, which can examine, copy, disrupt, or destroy data. If their progenitor's goal is only examination, you may never know they were there.

Once we have lost the privacy of our computers, it will be a natural extrapolation to surrender the privacy of our minds. The technology of mind mapping (as discussed in the *Immortality* section) is the foundation of a technology for examining our thoughts. The process will involve two keys steps. First will be the direct intrusion by molenorobots to do a mind map of our neurons and connectivity (as discussed in the *Immortality* section). This step could be eliminated if the information were already in storage. In the second step, AI will use the mind map information to synthesize our thoughts, which could then be e-mailed anywhere.

More advanced technology could reverse this process to replace our current thoughts with thoughts that would be more advantageous to the controllers (or other interfering organization).

In time such thought investigation and replacement activity will probably be legalized. This will be because through increasing sophistication of technology, a violent thought will have more immediacy; even a momentary ugly thought will pose a dangerous threat, since, without legal controls, the barriers between thought and consequent action will be reduced by the power of technology. Part of an over-all safety program will be to limit our use of potentially dangerous technologies; the other part will be to monitor and restructure our thoughts. Regardless of the legality of these activities, our vulnerability to them will be increased by encryption restrictions.

Also, referring back to the *Immortality* section, limited encryption will make more vulnerable the file that holds all the information to replace us (mind and body map) in the event of accidental death. This file can be covertly entered and manipulated. After our accidental death, on our subsequent replacement, we will no longer be who we were. We will be controlled by new ideas that were previously planted in our file.

Consider another example. Suppose you go to the hospital for major immortality refurbishment. The medical effort will be based on molenotechnology, and each molenorobot that enters your body to do repair work will be guided by its own computer. Each of these computers could be infected by a computer virus. Suppose that virus has been designed to instruct the molenorobot to enter various cells and mutate the DNA. Then after leaving the medical center, you age overnight and die.

A similar molenorobot could be designed to be put in a glass of water. After a person drinks it, it would do just as in the foregoing paragraph.

Almost everyone uses a car everyday. Suppose that in the future a person were to use a transporter. It is doubtful that average people will be allowed to use these freely, but this example is additional reason why people might not wish to use them. If the transporter's reassembly molenorobots were digitally contaminated in a specifically designed manner, the person would be reassembled with any designated motivation or mortal flaw. Doubtless, for health reasons everyone eventually would have protective sentry molenorobots in their bloodstreams, just as we have nature's white blood cells today, but these would occasionally fail to detect pernicious changes or be unable to deal with them rapidly enough.

Beyond any of the above concerns is the following. Once we all have molenorobots in our bloodstreams, they too will be controlled by whatever organizations are chartered with the responsibility for watching over us. These molenorobots will be susceptible to viruses designed by the watchdog organizations and any other capable organizations (or individuals) who wish to subvert the molenorobots so as to guide us to their own purposes.

(See also the end of the *New Nondigital Weapons* section. See also the *Communications* section and the *Digital Existence* section.)

5

THE RULING CLASS

The Shifting Elements of Power

Neither the elements of power, nor the ruling class that those elements sustain, have remained constant throughout the evolution of Homo sapiens societies.

In the jungle, for some simians, for example, leadership depends on some combination of physical strength, charisma, and intelligence. Voting can be seen even on this level when among the clan members there is a dispute about, for example, which direction to take. The voting occurs right on the ground with grunts, fists hitting earth, and quick eye movements about the circle of seated or pacing males. In such primitive political interactions, physical strength and intimidation have great immediate influence. After a few moments, the decision emerges.

As the ancient nations coalesced, personal physical strength was still helpful to winning a high place in society, especially since warriors fought with one another at a much more personal level than they do now or will in the future. On the other hand, in the more advanced civilizations, such as ancient Egypt, the highest positions—the pharaoh, nobles, and priests—usually came to depend more on superstitions, religion, traditions, bloodlines, posture, attitude, charisma, cleverness, and words. Certainly, too, material assets were fundamental to such rule, both by inherited position and as an attribute of commoners who rose to join the ruling class. However, at a certain level of a civilization's sophistication, assets that centuries earlier may have been accumulated through means that depended more on battle skills, are generally sustained and increased through the skills and structured bureaucracy similar to those of ancient Egypt.

This particular kind of ruling class lasted for thousands of years, and culminated in France in the combination of aristocracy and clergy just prior to the French Revolution in 1789.

With the invention of the printing press and onset of the industrial revolution, the elements of power again began to shift. Money and material assets began to flow more easily out of the hands of the aristocracies. Intelligence, cleverness, inventiveness, and charisma began to take a larger role in determining to whom money would flow. Along with the transfer of assets to entrepreneurs, went a rising influence over national activity. As the influence of monarchies decreased, the public politicians arose, bolstered by their own cleverness and charisma, and supported by wealthy people whose bidding they would do.

This has culminated in the U.S. and other developed nations, where the importance of public image and charisma has reached its apex through television. The successful politician finds a balance between the power of his own carefully orchestrated personal public appeal, and the deals he must make with moneyed interests to finance the advertisement and publication of his image; this is how votes are obtained.

Thus, today it is a collaboration and combination of charisma, clever maneuvering, posturing, and wealth that defines and sustains the ruling class. These are the elements of power in our society.

With advancing technology, the elements of power are again on the verge of shifting. The shift will be as dramatic as in the past and as rapid as the advance in technology. Wealth will continue as a key element, and it will shift to a new group, just as it flowed out of the hands of the aristocracy and into the hands of the entrepreneurs during the industrial revolution. Publicly promulgated truth will be important, as it is now, but there will be a change in who controls it. Advancing technology is the necessary backdrop for these changes, and the subsidiary driving issues are the increasing access to information and the capacity for its synthesis. Hence, the near future will summon the new ruling class—the controllers.

Witch Doctors, Kings, Priests, Presidents, and Other Liars

Truth, as perceived by the average person, is dictated by the ruling class, until they are deposed. They are the ones who have historically had the best and most timely grasp of the relevant truths, been most capable of altering the facts, had the best oratory (and writers), and been sufficiently in control of communications to promulgate their advantageous (self-serving) version of reality.

In an article by Lance Williams, Willie Brown, currently the mayor of San Francisco, was clearly revealed as the quintessential liar-leader. With regard to his record on tobacco industry issues prior to his mayoral election campaign, Brown, through apparent great intelligence and skill, was able to fill and refill any portion of the information void that remained on successive investigative revelations of the facts. The image that comes to mind from this smartly written article is that of a masterful man able to generate a tide out of his undulating version of truth, with which he can figuratively fill the San Francisco Bay; here and there a rock or island stands firm or emerges, but the tide slips around the significant rocks and, wherever possible, it rushes over the rest.

As a people, as a society, we demand that our leaders look good. We are an image-based, label-oriented society, always more concerned with how we look, what car we drive, what labels we may apply to ourselves or that might be applied to us, than who we truly are at the soul or spiritual level. Our leaders simply supply us with their own best image of themselves, just as we do for those around us. (Peer pressure is our premier button, despite our admonishment of youth to eschew it.)

Compared with today, in prehistoric times, truth was generally more immediate and obvious, since the fundamental relationship between provision and survival was more immediate: the hunt was successful, here is the carcass, let's eat. It was all very simple.

Gradually, spiritual leaders began to control some portion of truth. Their opportunities were enhanced by the large influence of unexplained natural phenomena, such as floods and epidemics, on the lives of their tribe's members. The witch doctors emerged to assuage fear of the unknown, and validated their positions with nontestable claims. And probably their assurance was a wonderful psychological panacea in times of turmoil caused by natural events, about which nothing could be done anyway. On the other hand, as abuses arose in their practice, our intelligence seeded suspicion. Still, in our ignorance of science and fear for our lives, how many of us would have dared to contest a witch doctor's claim of responsibility for the last solar eclipse?

Eventually, as farming, civilization, and royalty arose, truth fell into the hands of priests and kings.

Some religious leadership has been particularly abusive in its behavior, given its usual self-proclaimed awareness of spiritual truth. The history of some religions, and their involvement with politics, corporal power, material assets, war, witch hunts, and torture, suggests there is no special correlation between the clergy and spiritual truth, beyond what might be found among average lay people. The comforting core of each religion is a nontestable dogma that offers its devotees assurance of a special place with God (now or later). This assurance intentionally or inadvertently encourages attitudes and feelings of superiority, disrespect, and consequent rape, plunder, and murder (RPM).

The clergy itself is either naturally drawn to theological dogma or trained to its truth; subsequently, the dogma provides a vehicle for something far more important to them—their self-promotion. The fundamental problem, as with most leadership, is that the elixir of power usually takes precedence over fulfillment of promises to the constituency or adherence to spiritual values.

Gradually, kings took an increasing hand in creation of truth. Once the monarchies were deposed, truth became subject to the corporate and political leaders' advantages, though religious leaders today still play a role.

There is some reason to believe that scientists are more honest, since their work tends to fall into a realm that is more formulaic and more observable. This suggests that objective judgement will encourage more honesty. This is probably true, but scientists are human and subject to the same transgressions as other leaders when the opportunity presents itself. Examples of well-known respected scientists who are suspected of major fraud are Johannes Kepler, Albert Einstein, and Sigmund Freud.

Johannes Kepler (1571-1630) was the man whose destiny it was to validate Nicholas Copernicus' (1473-1543) heliocentric theory of the solar system, in the process of which he was the discoverer of the three laws of planetary motion. In 1609, in his certainty and eagerness to prove these laws to his peers, he used these (then) hypothetical laws to compute positions of the planet Mars. Then he presented this data as observations that proved the laws. This fraud was neither challenged nor discovered by his peers. It turned out that Kepler's laws were correct; indeed, they were the forerunners of Newton's simplifying and more encompassing law of universal gravitation. (Broad)

Albert Einstein (1879-1955) was the lover of a fellow physics student, Mileva Maric, a Serbian woman. At the turn of the century, for a woman to have wanted to be a physicist was very unusual; one can guess she must have been unusually brilliant, especially since she was a student of Zurich's Swiss Federal Institute of Technology, described by Overbye as the "MIT of Central Europe." In a love letter to Mileva in 1901 Einstein wrote, "How happy and proud I will be when the two of us together will have brought our work on the relative motion (relativity) to a victorious conclusion!" [Permission granted by the Albert Einstein Archives, The Jewish National & University Library, The Hebrew University of Jerusalem, Israel.] In 1901 they married.

Maric's Yugoslavian biography reports that the deceased Russian physicist Abram Joffe saw the signature Einstein-Maric on the original 1905 papers on the special relativity theory. This theory encountered great opposition when presented by Einstein (as his own). One can surmise that it was with mutual consent that the name Einstein-Maric was replaced with Einstein for publication. The social prejudices of that era probably made this choice seem wise. They probably thought they were serving the best interest of humanity. In hindsight, from a purely scientific view, given how much opposition the theory faced, perhaps some would argue that the best choice was made. Whether or not that is so, apparently fraud was committed by Einstein and his wife (Overbye).

Sigmund Freud, too, almost certainly committed fraud and malpractice. Frank Sulloway, author of *Freud, Biologist of the Mind,* an historian of science, and a visiting scholar at the Center for Advanced Study in the Behavioral Sciences, Stanford, California, says, "But the more detail you learn about each case, the stronger the image becomes of Freud twisting the facts to fit his theory." Regarding his inducement of Horace Frink and Angelika Bijur to divorce their spouses and to marry each other, Freud later wrote to Frink that he had asked Mrs. Bijur not to tell anyone of his advice that "she marry you on the threat of a nervous breakdown." In that letter he continued, "It gives them (people) a false idea of the kind of advice that is compatible with analysis and is very likely to be used against analysis" (Sulloway). (See also the reference to the sexual abuse case in the *Blindness* section.)

Words are miraculous in their power for enabling the articulation of truth, belief, claims, falsehoods, and lies. It is equally amazing that the clever speaker can create a false conception in the audience's mind without ever stating a falsehood.

An ancient Egyptian text of King Gheti's will to his son, about 2000 B.C., says, "Be an artist of words and conversation; the strength of man lies in his tongue; words conquer more than any battle" (quoted from Hatem). This is a wonderful, liberating truth, except that the words of successful leaders may contain too little truth. The ancient Roman playwright Terrence (circa. 186-159 B.C. wrote, "There's a demand today for men who can make wrong seem right" (quoted from Maddocks).

These ancient thoughts remain valid, and the associated transgressions will continue, but those whose greatest power is in their words will probably become buffoons in the near future. Those who have the wit to cleverly use words to their advantage will need additional technological skills to join the echelons of the elite rulers of the future.

Promulgation of Truth

Truth and falsehood are promulgated through communication.

Control of the means of communication has always been an essential element of political, economic, and military power, and hence of the control of a nation. This will continue to be so, regardless of the new form "nations" take. (As previously mentioned, geography will cease to be a determining issue; united groups will coalesce within intangible and digital boundaries.)

In ancient Egypt, the Nile River was the central means of communication for thousands of years. Oar and sail driven vessels plied its waters with people, chattel, missives, grain, and stone blocks for building temples and pyramids. Additionally, a postal road ran along the bank. (As mentioned elsewhere, the difficulty of moving men, material, and information across its borders, likewise kept invaders largely at bay for most of several millennia.)

In our evolving civilization of advancing communications technology, the most important means of communication are becoming (have become) digital. Thus, whether the medium is electronic or photonic, the increasing portion of our communications and information that flow through them implies an increase in the speed of our interactions, events, and our response to them. It consequently also implies a closer relationship between us and authority, and implies the necessity for more rapid assessment and a swifter response.

The New Ruling Class—The Controllers

A new ruling class, the controllers, is emerging.

Due to rapidly changing technology, its complexity, the immediacy of the impact of change, and its increasingly dominant role in our daily survival, these controllers are the ones who best understand our global technological, economic, and social reality. They will be the ones who have the required skill to rely necessarily and increasingly on computers and global communications to synthesize some meaningful image of worldly reality and the balance of competing and threatening forces. Comprehending this reality will be essential to meeting daily challenges to our and their survival as individuals and corporations. We will have no choice other than to rely on them.

Most current political leaders and most people generally will have neither the patience, intelligence, nor aptitude for sufficient training to understand truly what is happening, any more than they could learn to comprehend the formulas of general relativity. Political leaders, who generally lack sufficient ability, will be shut out of the truth generation process. Simultaneously, the controllers, together with the support of AI, will probably create fictitious digital political Internet personas that will be more witty, beautiful, and slick than real politicians. Thus, the power of politicians will fade.

Like the First and Second estates of France prior to 1789, politicians, as we currently perceive them, are blind to their imminent demise, and to ours perhaps so are we.

The controllers, being most clear in their understanding of reality, will also be in the best position to reshape truth in whatever way their strategic advantage dictates. This reshaping will be hard for anyone to challenge, or to credibly disprove, without very high intelligence and access to the same level of communications and computer support. Indeed, few of us will have the necessary skill even if we happened by some stroke of luck to get access to the necessary computer/communication tools. Thus, though information will become more accessible to the average person, determination of its veracity (or relevance) will be much more remote, and it will be relatively more difficult to synthesize.

These controllers also will be in control of the media, which will deliver their reshaped truths.

Thus, they will have much greater control over our perception of reality, as it applies to our survival, than any previous ruling class has had.

These changes will be coming soon. The controllers are beginning to rise.

6

CABALS

One fine day, long before the union of the Hawaiian Islands, on Maui's shore stood a good king, facing the island of Lanai and imagining himself to be a peace-loving king. The sunny beach was beautiful, the sparkling blue sea was fun to surf and full of food, the blue sky was clear except for white clouds that clung to his island's mountain tops to deliver an endless source of fresh water. In the midst of this happy perfection and abundance this king imagined a fellow king standing on Lanai's opposing shore looking across at his island. Knowing the power driven vanities of his fellow man, he wondered if the other king might be thinking of conquest. It made him happy to see his people laugh, play, and live their carefree island life. Above all else he wanted them to remain free. On this basis, though he imagined himself as peace-loving, he thought it might be preferable to preemptively conquer Lanai to end the fears he had for his peace-loving people. In his imagination, carrying out his royal responsibilities to his people, he mounted an invasion of Lanai and succeeded in conquering it, and he was relieved. (a parable)

Origins

The demise of the present ruling class will not be sudden. They will sense the increasing power of the controllers; they will sense their own failing power. Nonetheless, they would never publicly acknowledge that their power is fading; likewise, they would be reluctant to say they want to retain their power. They know that such admissions would clearly indicate their conflict of interest, thereby creating doubt about their suitability for subsequent leadership.

The present ruling class will gradually feel threatened; they will feel jealousy, suspicion, and fear. Then they will use their fading power to raise our suspicions.

The focus of the consequent public pressure will be to attempt to radically increase government control over corporations, the bosoms of which will be the heartland of the controllers. Through proposed laws, the government will try to make accessible to its agents all of each corporation's proprietary information (designs, methods, processes, etc.), just as it is doing with its increased interest in invading all communications (see *The Digital Battleground* section for government efforts to restrict the use encryption). Due to the advances in technology, this will be the main hope the political portion of the present ruling class will have of retaining its grip.

To whip up public support for such measures, the political leaders will have ample examples of terrorist and individual insane actions to ignite our fear about the dangers of advancing technology that is not under close government scrutiny or in government hands. They will falsely portray government hands as being equivalent to our hands, though necessarily much of this technology cannot safely be made public.

Many corporations, especially those involved in advanced technologies, already have large bodies of proprietary information and associated secure development divisions. The information security mechanisms are already in place and evolving. These corporations have cultures that traditionally recognize the importance of proprietary information for their competitive advantage, control, and survival. Also, they know that the government cannot nearly as well protect various proprietary information for them; through government interference, corporations would be more exposed to digital invasion by one another.

The foregoing, together with corporate awareness or suspicion that in fact jealousy and self-interest drive the attempts by governments to legally invade the corporations' intellectual assets, inner data sanctums, and libraries, will cause the corporations to react with hostility. They will feel justified in unlawfully withholding more data, plans, methodologies, and tools from governments and us, and in keeping their most potent weapons of covert warfare and other critical technologies from government oversight, and safely encrypted with advanced illicit codes. It is incidental that they and the government will agree on the danger of letting some tools and technologies become public.

As mentioned, corporations usually have proprietary assets, activities, and plans that are guarded by security personnel and equipment. Furthermore, various groups of corporate leaders naturally meet privately to discuss mergers, partnerships, technology, and world issues. Thus, the formation of a cabal does not have to be a conscious decision. Indeed, the members of such a cabal as discussed here could honestly deny their cabal's existence. The question really is this: "Given that these cabalistic groups wish to survive, at what point will they recognize the potential suggested in this book and begin developing plans to survive this future?" It will become clear that they cannot survive legally. It may only be at that point that some such organizations will first realize that they constitute a cabal.

Neither the ideas of this section nor of the rest of the book have to be completely valid, nor do the thoughts have to make complete sense, nor do all the scientific extrapolations have to come true, for dangerous circumstances to evolve. If only some powerful controllers come to believe in some of the foregoing, cabals will evolve and dangerous paths will be chosen. Even the worry that some other controllers might themselves form cabals, will motivate a suspicious group to create their own cabal as a necessary defensive measure.

Thus, within and between corporations, will form cabals—common interest groups with potent defensive and offensive tools. The cabals will arise quite naturally out of an instinct for survival. A typical intercorporate cabal might have members from a telco, a computer/communications manufacturer, an AI company, a digital weapons corporation, and a molenotechnology corporation. Out of a justifiable concern for economic survival and their members' physical safety, such combinations will make good sense in an increasingly technology dependent, complex, rapidly changing world.

The cabals' members will be thinking, caring people with families, perception, foresight, and the power to maximize their advantage in, if not fully comprehend and control, future developments. They will recognize that the rapidly advancing technologies (molenotechnology, AI, automation, and communications) have the significant previously discussed implications: (1) There will be much more rapid shifts in the balance of power between controlling entities; significant shifts may occur within minutes or hours instead of centuries; (2) small or poor entities will have an increased capability for negative, terroristic, or other controlling global impact; (3) the controlling competing corporations, cabals, and common interest groups will no longer include geographically based nations; and (4) the cabals will also perceive that there will be a rapid vast decrease in the size of the population necessary to support the further advancement of technology and their civilization as they and their families imagine it.

These things the evolving cabals will see. They will know they are in a survival game, and as human groups have always done, they will do whatever they think they must do to survive. What they may not anticipate is the consequent, potentially mortal, global implications of the combination of processes they implement, the weapons they create, and our decreasing importance and shrinking freedoms (as discussed in the section *Freedoms Willingly Surrendered*).

Once cabals begin to form, their existence and growth will be further encouraged in reaction to one another. Though their origination will have been defensive, they will become aggressive. They will carry on covert warfare with one another, using silent and publicly unobtrusive digital and molenotechnology weapons. They will digitally commit sabotage, deletion, theft, and falsification of data, just as corporations are already doing to one another, and will use offensive molenotechnology weapons. Such warfare will be at first covert, compared to weapons with which we are now familiar; it will become overt after it becomes the common manner of warfare, and after we learn to fear the global significance of digital destruction. These cabal actions will be against their corporate and cabalistic rivals. Additionally, as a minimum against the government, they will carry out preemptive digital strikes aimed at various invasive offensive threatening (though possibly legal) digital capabilities.

As previously mentioned, the cabals will be driven by an instinct for survival, and they will also have a mission—to save civilization. The alignment of these two drives will give them a powerful motivation.

The Need for Secrecy

The secrecy surrounding the existence of these cabals, or secrecy at least about their deeper concerns, fears, interests, speculations, plans, and purposes, is important to their survival and success. Individually and in alliances the cabals will fight one another, but they will have an understood common interest guarding the secrecy of their existence from us. If necessary, they will cooperate to maintain it.

Generally, the cabals will wish to work freely, without oversight from the public or government, and without creating suspicion or fear. Thus, regarding some issues, there will be a special need for utmost secrecy.

Obviously, the evolving uselessness of most of us is not a concept that a cabal would want to publicize, no matter how firmly they believe it. We would be very suspicious of any organization that tells us there will be no use for us in the near future. Rather, they need to prepare for the implications, whose startling nature demands secrecy.

Each cabal, recognizing the importance of secrecy, will embark on creation of tools, using advanced technology, to help it maintain secrecy. Also, it will need to watch for people who might create an early warning, and will possibly terminate them. This direction will itself circularly create an even greater need for secrecy.

Fear for the Perceptive Person

Anyone who surmises the future of things as discussed here, for a future so near, is a threat to the cabals, and therefore is at risk if such cabals already exist. Such a person would be well advised to keep these ideas hidden until they have published a convincing document, or else they may become the terminatable target of a cabal. If no cabal has yet formed, then the immediate danger is less; on the other hand, the danger could grow suddenly as soon as a cabal forms.

If no cabal forms until sufficient numbers of people are convinced of at least the possibility of a devastating future at the hands of the controllers or cabals, then the subsequent cabals will be less able to take action against single individuals in a campaign to stop the spreading awareness of this possibility. On the other hand, as long as almost everyone would deny the possibility of such a dire future, a cabal would probably be willing to murder anyone who seemed to have perceived the horrifying course of events that will occur within the next ten years.

Overt murder would not be the cabals' only option, and certainly they will have every reason to avoid martyring a seer. With the sophisticated tools available to them through molenotechnology, a person who appears threatening could be framed for any crime or given a traceless deadly designer disease; regardless of anyone's possible suspicion of foul play, nothing could be proven against the culprits, even if the microbe were identified. A person of great insight, but limited access to the tools of molenotechnology, will be helpless against the carefully focussed malevolent forces of the future.

Once a cabal uses its advanced technology tools to silence any single individual whom it finds dangerous to its existence, it will have a very high stake in secrecy surrounding the specific technology that allowed that perpetration, so as to not be suspected or implicated.

Danger to Us

Even if the goodness of Homo sapiens, as represented in the present and future ruling classes, is much greater than suggested here, the existence of only one malevolent cabal will make our existence precarious. If one cabal begins to think of world conquest, some of the other cabals will ally against it. In the subsequent war, which could be quite brief, but in which the digital and molenotechnology cross-fire will be intense, less protection will be afforded us due to our decreasing usefulness to civilization. Thus, many of us may simply perish as incidental casualties in a war for domination.

7

BLINDNESS

In the Holy Bible is found one of the earliest written acknowledgements of the propensity of humans to violate their greater spiritual light, and to rationalize their actions for doing so.

In the Garden of Eden, God tells Adam and Eve not to eat the fruit of the "tree of knowledge." They do so anyway. Later, God asks Adam whether he violated the dictum. Adam answers: "The woman whom thou gavest to be with me, she gave me [fruit] of the tree, and I did eat" (*Genesis* 3:12, King James Version). Adam stands next to God, his creator, the being who knows him best, and the one being who has the best possibility of offering compassion, understanding, and forgiveness, and the one who will certainly see through any falsehood or rationalization. This is the best chance Adam, or anyone, will ever have to be totally honest, to not rationalize, to take full responsibility. Adam instead implies culpability in God and Eve: "The woman, whom thou gavest," and "she gave me [fruit]." On the positive side, Adam does in fact confess: "I did eat." The difficulty is that such an unavoidable confession in the presence of the divine omniscient creator provides little to admire. By comparison, given the context, the impulse to rationalize is noteworthy.

Whether this story is factual or a parable, verbal recognition of a fundamental fault or deficiency in humans has never been so clear.

No matter how small or serious the offense and rationalization, we may be failing to see that on the allegorical level, parallel to that on which we lost our way in the Garden of Eden, and on which we lose more of it through our historical expanding desire for more material well-being, and in analogy with the continual carving of Earth, we are bartering away our souls, albeit only a little bit at a time.

The truth of this may not be fully comprehended by the average person until the first person has been entirely disassembled, uniquely stored in a data file, and then deleted. Of course, the possession of such incredible technology need not of itself imply the completion of the long process of self-annihilation of Homo sapiens' soul; rather, this implication comes from the manner of our acquisition of such technology and at whose expense.

The Power of Pleasure

Pleasure has many different aspects, but it generally refers to sensual, physiological, emotional, and spiritual delights; certainly too a feeling of pleasure usually accompanies the acquisition of cash and other forms of increase in our material well-being, or the increase in our comfort. All of these generally constitute "enhancements" to survival.

Both ancient and modern thinkers have recognized that our beliefs become biased by benefits.

In Deuteronomy 16:19, Moses addresses priests:

> A gift doth blind the eyes of the wise, and pervert the words of the righteous (KJV).

This truth applies to gifts already received and to anticipated benefits; either kind molds our thinking. Consider the profound implication by the vice president and editor-in-chief of *Variety* in the following quote:

> Why is Wall Street now telling the big entertainment companies that unless they spin off ancillary operations, their stock prices will continue to be depressed? Only a couple of years ago, these same advisers were preaching the doctrine of vertical integration—it was mandatory to control all facets of production and distribution. The investment bankers, of course,

make their money on deals. Now that these megacompanies have bought everything in sight, could it be the bankers would now like them to sell everything in sight? (Bart)

While these two examples refer mainly to the trade of our own benefit to the detriment of others, an equally potent human characteristic is that, individually and collectively, we are willing to knowingly trade our long-term survival for immediate enhancements. This trait is well-known; there are many examples. Probably we do it because it adds to our enjoyment of life.

For example, to most people who smoke, smoking is an enhancement—they enjoy it. Many of these people are fully aware of scientific evidence that proves the correlation between smoking and the increased risks of disease and earlier death. Though many humans have chosen not to smoke, the trait exemplified by smoking is common to us all.

All of us live in various degrees of pleasure, wellness, sickness, and pain. All drugs, for example, have toxic side effects, yet most of us take them for increased pleasure, cure, or relief from pain, and that choice is similar to the choice to smoke.

By way of analogy, on a global level where our accumulated individual choices have destructive global side effects, is the choice to smoke any different than the choice to use cars, trains, airplanes, and boats? The manufacture of these, and our consumption of fuel to run them, creates an equivalent set of deadly pollutants that we sooner or later breathe, drink, and eat. A choice is made each time we commute or travel. Regardless of how totally this choice is ingrained in our society, and how much we think we must use such transportation for our livelihoods, the truth is that this choice simply sustains a standard of living and its associated pleasures. That we now take these for granted is why raising the question of it, or pointing out that it is a choice, is apt to be met with indignation or ridicule. Nonetheless, self-righteousness cannot negate the fact that pleasures and comforts are being traded against detrimental side effects. Yet we do it.

Likewise, the choice to smoke is similar to the choice to eat unhealthy food. We may enjoy it, but it is bad for us. This issue was discussed by the ancient Greek philosopher Socrates (470-399 B.C.) in his dialogues. Of course, he was put to death for his ideas. Though he wasn't put to death for his dialogue about unhealthy food, it is ironic that he chose to drink hemlock to meet the requirement of his execution.

Sure, there are risks in a chocolate sundae, skiing, or surfing, but these things are fun, especially when there is a rush of adrenalin to boot. Life is boring when we choose to take no risks. Enhancements add spice to life, even if, and sometimes because, there are risks. Similarly, in a passionate moment we may choose unsafe sex, a risk most of us at least occasionally have taken. And enhancements that add to our comfort are almost always welcome, especially when the risks are hidden or distant enough to ignore.

This is very similar to the trait that gives higher priority to immediate problems or dangers, than to long-term ones (as discussed in the *Natural Environment* section); indeed, it probably originates in the same survival mechanism in the primitive depths of our brains. It is so deeply ingrained in us that for immediate benefits we are willing to pay a much higher price later.

In comparison, it should not be difficult to admit that we are even more willing to make the trade if the price will fall on the shoulders of a later generation. Likewise, it is easy to understand our readiness to trade on the declining well-being of other life. Certainly, it should be no surprise that if we are willing to trade immediate benefits against ourselves individually, we are also willing to trade them against the well-being of others.

For each of us there is a tendency to dissociate from whatever is not ourself now. Whether it is ourself later, another person, or any other being, the natural survival mechanism in our brain is biased in favor of the present self. (Generally, motherhood is an exception to this.) This deeply ingrained survival instinct causes effects that spill over into nonsurvival areas of our lives and inhibits our spirit from protectively encompassing all others as fully as we serve our immediate selves.

Thus, armed with the will and the power, and loaded with

such self-interest, almost by definition aggression coagulates in every conceivable form, except for the ironic frequent exclusion of the name.

There are particular additional pleasures that accrue to the ruling class through wealth or political power, not the least of which are felt in the status of having what others have not. These pleasures motivate many of us. Of course, the relative nature of this claim is best comprehended through comparisons between various groups. By such comparisons we can see that always, with respect to some group, we have that which with respect to another we claim to have not. And with respect to those who have less than ourselves and less power to resist, our behavior varies little from that of the ruling class, about whom we frequently grumble.

Wealth is always relative. There are many ways in which we recognize this: (1) what we have on the average in today's world compared to what people of past generations had, (2) what we have on the average today compared to what we had in the past, (3) what we have with respect to one another individually, (4) what the ruling class of our nation has with respect us, (5) what we as a nation have with respect to other nations, and (6) what the some half billion of we citizens individually have in the modern nations, and in the comparably affluent pockets of wealth within the impoverished nations scattered across the globe, compared to the average people of those impoverished nations.

As we approach the borderline of human starvation and death by disease, even the most unfortunate of humans at least theoretically has a greater proclaimed right to life than any other animals, regardless of the many cases of contrary practical reality. Exceptions are of course found in this basic commodity of life. Consider the imminent death itself and the relative physical comfort of living persons who are dying because they have so little nutritious food compared to, for example, a deer who lives healthy, wild, and free in an abundant uncut forest, or the the well-fed thoroughbred being prepared for a race.

Political power—power over other beings' lives—too is relative. Nonetheless, it is also by definition a limited commodity. There may be limitations between any two groups as to how much political power is available in their struggles with one another; however, if we compare ourselves to all possible groups, we see that the meaning of political power and whether we possess any, is a relative question—one that can always be answered affirmatively.

Being careful not to confuse two distinct meanings of power—political power and the motivational power of sought-for pleasures—we may consider pleasures due to political power. Political power can give a great deal of pleasure.

Henry Kissinger, former secretary of state to Presidents Nixon and Ford, stated this:

> Power is the best aphrodesiac.

This is not the first expression of the connection between control and sexuality. Psychiatrists who study rapist/serial killer personalities are familiar with this connection too. Nonetheless, Kissinger might be commended for such candor, which is unusual for a person in a political position.

Equally startling and broad in its implication is the following quote of Adolf Hitler (1889-1945):

> It also gives us very special, secret pleasure to see how unaware the people around us are of what is really happening to them (Fest quoting Hitler).

These quotes are included for their exemplary nature. That there is great pleasure in political power is commonly known and accepted.

One may wonder about the extent to which the possession of political power creates pleasure that is similar to that of heroin addicts or street junkies, and whether it is equally addictive, though it does not seem to be as physiologically destructive to the user. It is difficult to judge which drug creates the greater motivation; this is partly because the manner of supporting the habit is so necessarily different, as is application of the dosage, which for politicians is offered in four- or six-year supplies.

In comparison, some financiers on Wall Street have said that they have become addicted to money, and that they have begun to feel they just can't get enough.

In the case of narcotics, the habit is more apt to be supported by lying, prostitution, and theft. While some plaintive thinkers have stated that these also are the stock and trade of politicians, the activity is definitely white-collar and deemed respectable; in any case, the result is "good" leadership, which is thought to have socially redeeming value.

Coordinated aggressive societies demand leadership, and hence a ruling class. To eschew leadership is to guarantee being overrun by a "superior" (more powerful) society. To accept leadership, indeed to support it, gives one's own society the greatest chance to be that "superior" society, and to enjoy the resulting benefits in which our survival is made easier or its enhancements less expensive.

In the movie *The Immigrants,* with the Swedish actor Max Von Sydow, one of the settlers in Minnesota complains to the character played by Von Sydow about his fears of the restless Indians on the recently established neighboring reservation. Von Sydow's character responds with comments about how hungry the Indians are since it is winter and they have lost their lands to the white man. The other settler defensively replies, "I paid good money for my land." Von Sydow suggests that the price was very low. As always, stolen goods are cheaper, no matter how dear the price may seem to the purchaser.

So property is taken and secured by soldiers, by society's aggressive action arm. And then, once someone (a civilian) has purchased or occupied the stolen property, that person becomes financially and emotionally vested. He is ready to self-righteously fight to keep it, even from its rightful owner.

Simply by not questioning the property price, by abdicating our responsibility to do so, those of us physically or temporally behind the front lines become willing, emotionally committed participants in the RPM process. We do it out of our desire to have better, more comfortable lives, even as we may beg the question of our culpability, by our complaints of the price and how hard we have to work for it. So it was as our ancestors stole from the Native Americans, and so it is today as we take the rest of Earth for ourselves.

We allow ourselves to be blinded by the benefits that our leadership promises and sometimes delivers. And they rightly see it as their job to help us ignore what they know we truly do not wish to see. Who is to blame if the ruling class creates delusional dreams and happy fantasies for us? That is what we want them to do, and we reward them by buying their products and giving them our votes. Thus, in bootstrap fashion the propensity for leadership has created a need for itself. On the other hand, in our own willingness to be repeatedly lied to and defrauded with promises, there are increasing signs that suggest we have followed a path into a quagmire of death.

Perhaps we had to come this far to see where it might lead. Probably it is irrelevant that we have never been anything more than pawns in the pleasure power games of the ruling class. Certainly they did not (and do not) know the endpoint any better than we; their highest priority has always been simply to stay in control, to retain their superior position. Even if they were not blinded by their own power-pleasure habit, even if their addiction were eliminated from the loop, the future, even that which has already past, is usually difficult to know. The most honest of leaders have always meant well, and often so have the least honest of them. And so it is with the rest of us.

Denial

Strength and Weakness

We have an effective tool for survival and its enhancements: It is called denial. Denial adds to our blindness about certain things, and usually this blindness increases our probability for survival (regardless of whose expense).

Denial comes in many forms and is promoted by many things. At its heart is the issue of truth and belief, and how human belief is biased in the face of (1) a desire to improve the chance of immediate survivability, (2) desire to promote the stability of society, (3) anger, (4) wanting something, (5) wanting comfort, (6) praise or a desire to hear good things about ourselves, (7) fear, etc.

(1) Desire to improve the chance of immediate survivability

A fundamental survivalist rule is to maintain a positive outlook in any situation. In many circumstances, believing that you will survive increases your probability for surviving. Graduates of Outward Bound know this as well as does James Bond.

On the other hand, Japanese leadership at the end of WWII had become so firmly entrenched in their hope for victory, that the U.S. needed an atomic bomb explosion to convince them that they had lost the war.

(2) Desire to promote the stability of society

The stability of society is often given an importance that overrides truth. Examples range from Torquemada to Galileo, and beyond.

Tomas de Torquemada died in 1498. He was the Spanish grand inquisitor. Given the horrors of the inquisition, perhaps his death date should be celebrated as a holiday. Of course, for all its terrible reputation, the inquisition may have supplied a stabilizing force to society. Certainly the views that it tried to suppress were uncomfortable to the catholic authorities. They doubtless were worried about their own importance within society; consequently, they would have worried about the solidarity of catholicism and its ancillary stabilizing role.

Similarly, in Italy the church threatened torture to force Galileo (1564-1642) to deny that Earth moves in an orbit around the sun (Chambers, *et. al.*). Galileo's scientific observations felt threatening to the church. On his way out of the building after this incredible moment of denial in human history, Galileo commented, "Nonetheless, it moves." At the moment of this utterance, he made sure he was staring at the wagging tale of a dog.

(3) Anger

Relating to the above examples regarding the importance of the stability of society in biasing determinations of truth, the beautiful Greek philosopher/mathematician Hypatia also serves for an example of how anger can justify almost any action, regardless of one's religious teaching.

> Hypatia...was the first notable woman in mathematics.... became the recognized head of the Neoplatonist school of philosophy at Alexandria, and her eloquence, modesty, and beauty, combined with her remarkable intellectual gifts attracted a large number of pupils.... Hypatia symbolized learning and science, which at that time in Western history were largely identified by the early Christians with paganism. As such, she was a focal point in the tension and riots between Christians and non-Christians that more than once racked Alexandria (from "Hypatia" in *Encyclopaedia Britannica*, 15th edition, 1992, 6:200).

Beyond Hypatia's non-Christian beliefs, several things made her a likely target. She was a friend of Orestes, the pagan prefect of Egypt, and there were scandalous stories about the nature of their relationship. Furthermore, she was said to support his political opposition to St. Cyril, patriarch of Alexandria. (*Encyclopedia Americana*)

> In March 415 (A.D.) a Christian mob in Alexandria, incited by fanatical clergy, stopped her carriage one day, carried her into the Church of the Caesareum, tore her limb from limb, and burned her broken body in the street. (From the *Encyclopedia Americana*, 1997 Edition. Copyright 1997 by Grolier Incorporated. Reprinted with permission.)

Probably they felt they had a good reason.

While the above examples relate to the issue of what people want, and how that determines what they believe, in many cases what we want is less overt or less immediate. For example, most of us rely on the stability of society for our long term survival, and gain great comfort from it. Stability is something we want, and the many ideas and beliefs that support it are shaped and evolve over periods of time. On the other hand, sometimes it is a more immediate want or need that shapes our thoughts and justification processes, and complete blindness may still result.

(4) Wanting something

The Puritans who landed at Plymouth Rock in 1620 viewed themselves as persecuted, peace-loving, God-fearing Christians. They only wanted a place to live in peace, and to worship without persecution. Yet they were the unwitting vanguard of one of history's most horrifyingly memorable incidents of RPM. (This of course would be discounting Homo sapiens' present daily multiple genocide of species.) That we do not usually have that view is because of our empathy for the Puritans, who were also our forebears, and because the RPM took place over a long period.

European Christians used mummy powder imported from Egypt as an internal medicine during the 1500s until the 1800s. This was blatant cannibalism. Additionally, it is ironic that in eating this ancient flesh, these people were thoughtlessly destroying, within the constructs of the religion of ancient Egypt, any hope of immortality for the corpses.

At a church service the minister praised man's vision. As an example, he praised man's vision in turning a desert at Palm Springs into a golf course. He mentioned how "there had been nothing there," and that men saw a golf course and drew water from the ground to make it green. To an environmentalist it is appalling that this minister could think nothing lived in the desert; it is equally appalling that he did not know that when we lower the water table under the desert to water our golf course, then elsewhere, perhaps many miles away, springs dry up, and the local animals die of thirst. With all man's vision for his benefits comes a blindness to some others' needs.

Likewise, to further the point, we increasingly fancy ourselves enlightened now, at least about whales. Now most of us are satisfied to only view or take pictures of them, instead of killing them. However, it is not good enough to simply stop hurling harpoons at whales; we must take a broader view. When we consider all the pollution associated with cameras and film, each photograph is like a poison dart fired at the subject, a toxic pill dumped into the ocean and spewed into the sky. Taking photographs of whales today is the equivalent of hurling harpoons at them 100 years ago. It may not be so immediately mortal, but in the context of modern society and toxic waste, it is just as blind, just as foolish, and in the long run just as deadly.

Yet, if scientific studies showed that within twenty years, pollution deriving from cameras, photographs, videocameras, and videocassettes would account for the death of all whales and half the remaining species on Earth, how many of us would believe such a thesis? Such an idea would be scorned and buried by film and camera companies. Even if they had no response, the majority of us would discount such a thesis for presumably logical reasons, little realizing that the true source of our doubt would lie in the inconvenience our credulity would impose.

Regardless, with but a little thought about the toxic impact of photography (and many human activities) it can be clear to anyone who wishes to see it, that in at least an allegorical sense, photographs steal the soul of the whale, and of the planet generally, and ultimately will kill them both. Native Americans and other aboriginal peoples have been right all along: despite our scornful mirth, photographs steal our souls.

With the rise of the Internet and e-mail, it is popular today to herald the evolution toward a paperless society as a step in favor of environmentalism. Whether or not it is, this idea suggests complete ignorance of the incredibly high toxic price our society has tithed the planet to achieve, reach, or stumble onto, and to continue the present state of technology. Fewer trees will be cut; however, there are less trees available to be cut since we have already cut and used them. Something much more deadly may be replacing the fall of trees; its relative silence gives less warning. To support the computer/communications technology of the paperless society we pay a large silent toxic environmental price. On the other hand, the immediate material benefits of being an electronic rather than a papered society are very obvious. Thus, to pride ourselves on this as an environmental achievement makes as much sense as it would to take pride in our moral qualities based on the fact that we are no longer slaughtering Native Americans, since we have already beaten them.

In an environmental report regarding the debate over the spotted owl and timberlands in the Pacific Northwest, one well-meaning lumberman, who doubtless was concerned about his job and his company's profits, sought to justify endless logging to the last tree by saying, "Bugs, fire or man are going to harvest the trees; they don't live forever" (Gup). How many of us are able to see, in the midst of concern for our jobs, and the need for money to feed our families, and to make our house mortgage payment, that if such reasoning were valid, then it would equally well justify Jack-the-Ripper?

(5) Wanting comfort

Our desire for comfort is a natural one, but it can wreak havoc with our intellectual capacity.

For example, reverse reasoning, though logically faulty, can enhance survival, at least by lowering our stress and increasing our comfort.

Michael Caine and Omar Sheriff starred in a little known film, *The Last Valley.* In it Sheriff's character says to Caine's character, "My 12-year old sister was a witch. She must have been. They tortured her and burned her at the stake." This important example of man's thinking is not only in the past or in movies. To legitimize the horrors we see around us, we sometimes will accept for truth that which makes us most comfortable. The horror of a little girl having been tortured and burned is less if we can believe in addition that she was an evil witch. By our nature many of us have a tendency to believe a false claim after a consequence has been imposed by authorities; this is especially so when we feel powerless to change it.

Slavery has been justified in many ways. An excellent presentation of the balance of moral, economic, and religious pressures surrounding this institution was given in the film, *The Mission*. The film stars Jeremy Irons and Robert De Niro, and the setting is South America in the 1700s. At that time slavery was proscribed by the Catholic church. The loophole was that the proscription applied only to humans.

As presented in the movie, the farms run by the Guarani indians under Jesuit guidance were commercially competitive with Spanish and Portuguese immigrants' farms. So, economic pressure was converted to political pressure on the church, which consequently effectively ruled that the Guarani were not humans. Thus, instead of having to compete with the natives, the immigrants were allowed to capture and enslave them—a diabolically poetic Christian solution. In the final poignant scene between the Portuguese ambassador, the Spanish ambassador, and the cardinal who was sent to provide the ruling, all parties understand in their hearts that the Guarani are human, and they know that the decision given by the church was for economic reasons. Furthermore, in the subsequent removal of the indians from the missions, there was a great deal of slaughter. The rationalizing Portuguese ambassador says, "The world is thus." The cardinal replies, "No señor, thus we have made the world" (*The Mission*).

Many people who saw this beautifully made film in a theater simply sat motionless in their seats after it was over. It was not popular. It is easy to see why. It was too true about something ugly, and thus too uncomfortable.

The First and Second estates of France were three percent of the population before the revolution. They were the nobility, the aristocrats, the priesthood. If they could have foreseen the revolution, they would have behaved differently. They were incapable of seeing it. It was easier, more comfortable, to imagine things would continue as they had for centuries.

Sometimes, in our circumstances of peer pressure and other recognizable and acceptable societal stabilizing forces, we are not able to see the truth, we are not allowed to see the truth, we deny truth. Few of us wish to be ostracized.

Freud was either a malevolently abusive fraud, or blind to his own support of a parent's encouragement of child molestation within his own case load: in the case of 16-year old Ida Bauer (Dora) it stared him in the face (Goleman). To give him the benefit of the doubt, we must assume he was blind. (See further Freud failures in *The Ruling Class* chapter.) But at his time and place in history, how would he have fared as a professional or as an individual if he had been able to see and say what he saw? This kind of peer and social pressure is implicit; a person is apt not to think of, consider, or give credence to an idea that, if promulgated, would bring certain professional demise, personal embarrassment, and shame.

Can we blame Freud for his bias? Are any of us any different? If we want others to feel safe around us, if we want to keep our job, we should not say certain things, regardless of their veracity. And what we should not say is correlated with what we should not think.

(6) Praise or a desire to hear good things about ourselves

Generally, humans, whether individually or in groups, or as a species, like to hear exciting, special things about themselves, things that set them apart and make them feel unique or more important. Most of us do, and often it is love or harmless fun that promotes such egocentric statements and our acceptance of them.

For excellent examples of this, the reader may refer back to the *AI* section for Michael Conrad's ludicrous comment, "Humans were the first computers." Also mentioned there is the *Life* magazine article about the human brain and beginning with the ridiculous claim that, "It is the most highly organized bit of matter in the universe."

As another example, it is common among Homo sapiens to feel that babies are born good. What does this idea mean, where does it come from?

Babies are born hungry, wanting, needing, demanding, and with no regard for other beings, and with very little control over their behavior or biological functions. While ignorance and lack of self-control may earn a baby our compassion and loving assistance, and provide a reason for its thoughtless behavior, and is common to baby beings of most species, it is a mistake to equate its condition with goodness, even if a baby's innocence is fundamental. On the other hand, our ability to cherish such a useless creature might appear to be one of our finest attributes, for it is based, through instinct and thought, on the hope for the future and survival of the species.

Also, we love them because they are helpless and because we are optimistic about their future growth. Despite our own failures and shortcomings, we see in them the potential of what they might become—what we might have become—and in this vision we overlook their temporary incapacity. On the other hand, when looking to the eventual adult grown from a baby, a realistic vision would acknowledge the potential for good or bad.

There is no benefit to assuming it will be good just because it is human, unless we acknowledge this assumption as a manifestation of another fundamental human character weakness—too frequently identifying that which is human as that which is good. But which parent can resist the temptation to identify a baby's birth as a miracle and sacred, given that doing so automatically brings we biological parents into the same divine circle?

It is difficult to assess the point at which myths about ourselves become dangers, but we are entering a period of technological advance and control in which a reality check on our strengths and weaknesses could be our only hope for survival.

(7) Fear

The course of a lynch mob, like that which tore Hypatia apart, is directed by anger and fear. Anger inspires it. People get so upset over a perceived or real injustice that they seek immediate retribution or vengeance. Angry people do not like to consider that they might be wrong. Thus, even before the mob has chosen its victim, it is hard for anyone to raise their voice against it, for fear of becoming the target of mob anger.

Humans have maligned sharks for millennia. People fear sharks. Most people would probably consider them to be the most vicious killer on the planet. In fact, humans kill one million sharks for every shark that bites a human (Manire). And humans surely kill many times the number of humans that sharks kill. To be sure, sharks kill lots of other animals, but so do humans, who exterminate hundreds of species per day.

Many rapes begin without threat of violence, even when murder will be the result. The predator moves in slowly, testing the potential victim's boundaries. For example, the perpetrator may ask for a cigarette light, and move especially close. Succeeding thus far, possibly then he will ask for money. The increasing proximity and pressure allow the perpetrator the chance to sense the intended victim's vulnerability.

No gender bias is intended here. Women rape and kill as well. However, usually for women these acts are carried out with more finesse, over longer periods of time, and with less violence—poisoning may be preferred over stabbing.

Once the rapist has taken control, and the victim knows that rape is intended, the rapist may demand that the victim get into his car. The now frightened victim begins to make mental trades: She may accept that the rape is going to happen and she only hopes that her life will be spared. As part of the bargaining that may occur in conjunction with her getting into the car, and thus giving herself more over to his control, she may demand assurance that he will not hurt (kill) her. He will give the necessary assurance, and she will get into the car.

Obviously, from a logical standpoint a perpetrator's promises are the last ones on which a victim should rely. Our gullibility can become especially high when we are in fear. Part of the reason we are inclined to accept the perpetrator's promises of relative safety (in return for giving him greater immediate control) is our fear of making him (or her) mad, and thus bringing to quick fruition our worst fears. The victim is bargaining for the best deal (promises) in the immediate moment, at the expense of certain horrible future moments and possible death.

By analogy, the ruling class behaves toward us in the same way as a rapist toward his victim. It finds the fears and insecurities, and relies upon the implicit threats to encourage the masses to allow an encroachment upon their space and freedoms. We behave like the victim, fearful for our immediate safety, willing to believe the promises of offered safety and benefit if the requested freedoms are to be sacrificed. We give in, hoping for safety, though we may still end up dead in the gutter.

Whether out of an ugly feeling of horror that humans could do such things, or a desire to believe only good things about ourselves, some people do not believe the Jewish Holocaust of WWII occurred. This blindness is the same as that which allowed people who lived within ten miles of a death camp to not know of it, and it is the same as that which makes people sure that such a thing will never happen again. If it indeed never occurred, then the horror would still be startling, since those who believe it did have put a lot of people to death for it. Regardless, the mere fact that such people exist and that others believe to the contrary, shows that somebody is in denial.

The truth is often denied when it will distract us or drain too much of our energy. Similarly, we are less apt to examine as carefully the veracity of a claim that fits what we need to believe, or would be to our advantage to believe. Generally we try to ignore truth when it will divert us from effective profitable action, or when it threatens to motivate us away from comfortable inaction. In these respects, denial has usually aided our survival (without considering the justice of it). But the Holocaust proves that this blindness has its ugly side, its dangerous side—dangerous to others and to ourselves.

Hitler's Holocaust

In Hitler's Germany, from 1933 through 1941 (and after), property, privileges, rights, and freedoms were gradually stripped from Jews. The following are examples:

April 4, 1933: Jews were barred from employment in civil service and all other public employment within the government and universities (Levin and *Britannica*).

September 15, 1935: The Nurnberg (Nuremberg) Laws deprived Jews of German citizenship, and forbade marriage and sexual relations between Jews and Aryans.

November 16, 1937: Jews could obtain passports for travel abroad only in special cases.

July 23, 1938: As of January 1, 1939, all Jews had to carry identification cards.

November 11, 1938: Jews could no longer own or bear arms.

December 3, 1938: Jews were required to hand in their driver's licenses and car registrations.

September 23, 1939: All Jews were required to hand in their radios to the police.

July 29, 1940: Jews could no longer have telephones.

July 31, 1941: The "final solution" began.

September 1, 1941: Jews could no longer leave their places of residence without permission of the police.

December 26, 1941: Jews could no longer use public telephones.

April 24, 1942: Jews were forbidden the use of public transportation.

October 24, 1942: All Jews still in concentration camps in Germany were to be transferred to Auschwitz (an extermination camp).

In September 1935 the appropriation of Jewish owned properties began. By 1937, one fourth of German Jews had emigrated (Levin); these doubtless found the discrimination too acute, and possibly too threatening. For context, *Kristallnacht* (Night of Broken Glass) occurred on 10-11 November 1938. By then all Jewish passports had been revoked; it was too late to emigrate.

Regardless of the emigrations, many Jews could not see the deadly implications of the changes until it was too late.

> Many believed, particularly in the early years of the regime, that it would be possible to work for a renaissance of Jewish

> culture in Hitlerized Germany.... German Jews experienced a new strength before the end, magnificent bursts of spiritual renewal on the eve of extinction. At the same time, this retrieval of old buried Jewish springs delayed recognition of the doom facing them (Levin).

Of those who stayed past the revocation of passports, most of them could not anticipate the Holocaust, for it was unthinkable, until they were being herded naked into the showers of death.

The following is a partial quote of Himmler's Posen Address to his SS group leaders in 1943, praising them for their good work:

> I shall speak to you here with all frankness about a very serious subject. We shall now discuss it absolutely openly among ourselves, nevertheless we shall never speak of it in public. I mean the evacuation of the Jews, the extermination of the Jewish people. It is one of those things which is easy to say: "The Jewish people are to be exterminated," says every party member. "That's clear, it's part of our program, elimination of the Jews, extermination, right, we'll do it." And then they all come along, the eighty million upstanding Germans, and each one has his decent Jew. Of course the others are swine, but this one is a first-class Jew. Of all those who talk like this, not one has watched [the actual extermination], not one has had the stomach for it. Most of you know what it means to see a hundred corpses lying together, five hundred, or a thousand. To have gone through this and yet—apart from a few exceptions, examples of human weakness—to have remained decent, this has made us hard. This is a glorious page in our history that has never been written and never shall be written.
>
> The wealth which they [the Jews] had, we have taken from them. I have issued a strict command...that this wealth is as a matter of course to be delivered in its entirety to the Reich. We have taken none of it for ourselves. Individuals who have violated this principle will be punished according to an order which I issued at the beginning and which warns: Anyone who takes so much as a mark shall die. A certain number of SS men—not very many—disobeyed this order and they will die, without mercy. We had the moral right, we had the duty to our own

> people, to kill this people that wanted to kill us. But we have no right to enrich ourselves by so much as a fur, a watch, a mark, or a cigarette, or anything else. In the last analysis, because we exterminated a bacillus we don't want to be infected by it and die. I shall never stand by and watch even the slightest spot of rot develop or establish itself here. Wherever it appears, we shall burn it out together. By and large, however, we can say that we have performed this most difficult task out of love for our people. And we have suffered no harm from it in our inner self, in our soul, in our character (Fest quoting Himmler).

This was Himmler's Holocaust thank you speech to his SS group leaders who were helping with the extermination of the Jews. There are no enlightened comments that could surpass the importance of a reread. It is enlightening, or frightening, to note the inclusion of the following concepts: decency of the SS group leaders (executioners), glory, ethical purity of execution of the program, moral right, duty, self-defense, love, testament to the wholeness of the inner self, soul, and character of the SS group leaders.

Himmler had a compassion for animals, some animals:

> How can you find pleasure, Herr Kersten, in shooting from behind cover at poor creatures grazing on the edge of a wood, innocent, defenseless, and unsuspecting? It's really pure murder. Nature is so marvelously beautiful and every animal has a right to live (Fest quoting Himmler).

At another time, Himmler said:

> One principle must be absolute for the SS man: we must be honest, decent, loyal, and comradely to those of our own blood and to no one else. What happens to the Russians, what happens to the Czechs, is a matter of utter indifference to me. Good blood like ours that we find among other nationalities we shall acquire for ourselves, if necessary by taking away the children and bringing them up among us. Whether the other nationalities live in comfort or perish of hunger interests me only insofar as we need them as slaves for our society; apart from that, it does not interest me. Whether or not 10,000 Rus-

> sian women collapse from exhaustion while digging a tank ditch interests me only insofar as it affects the completion of the tank ditch for Germany. We shall never be cruel or heartless when it is not necessary; that is clear. We Germans, who are the only people in the world who have a decent attitude toward animals, will also adopt a decent attitude toward these human animals, but it is a crime against our own blood to worry about them or to fill them with ideals (Fest quoting Himmler).

Today most people are ready to fight to prove they would not do what Hitler, Himmler, and Eichmann did. It is almost too horrible to speak of, much less to consider oneself a potential perpetrator. But what about being part of the compliant majority?

Possibly, most of us would do exactly what most Germans did if we found ourselves in their circumstances. For example, suppose in Hitler's Germany you are the father of a family that lives down the road from an interment camp, which you do not know is actually a death camp. You have a loving wife and two sweet teenage girls. You work in a nearby factory. Guards from the camp drink at the same tavern as that in which you have a beer after work each night; you hear things, monstrous things. These things are too horrible to think about; they can't be true; the guards meant something else. What can you do anyway?

Perhaps you are concerned enough to discuss the rumors with your wife. Being a caring sensitive woman, she will not believe such things could be happening. If you ask questions of the authorities you will be assured that nothing extraordinary is happening, and you will be told not to ask questions. There may be further discussion between you and your wife in the darkness of night after the children are in bed. Phrases like "loyalty to the state" and "leadership of the Fuhrer" will be solemnly mentioned.

If you publicly persist in your questions, you will lose your job. How will your dependent wife and daughters be fed then? You (a good citizen) would be failing them as a husband and father to persist in questions. Either the majority (fearful for themselves) or the minority will attempt to silence you. If you persist anyway, you will disappear without a trace.

Usually our own survival, and that of our family, comes first. Many of the pressures for conformity and obedience are covert in their implicitness, without being invisible.

In the inspiring and depressing film *Au Revoire, Les Enfantes,* the fathers of a catholic boys school give Jewish boys false (non-Jewish) names and hide them within the student body. Near the end of the film the ruse is unveiled. Finally, the Jewish boys are taken away by German soldiers, as are the school's compassionate fathers.

There were many Germans and others who risked their lives and their families to help the Jews. Doubtless, most of us today would believe that we would have done the same. We certainly would *like* to believe this about ourselves. It is almost unthinkable to believe otherwise; we could barely escape public scorn if we stood up in public and said we would not have had the courage to help the Jews.

At Humbolt State University (in California) Holocaust survivor Samuel Oliner and his wife Pearl did a study of those who helped the Jews during the Holocaust. He compiled a profile of the type of person and family who was willing to risk everything to help the Jews. We each might wish to see whether we fit the profile (King). We each might ask ourselves what type of people we really are, as opposed to how we would like to believe we would behave in a test that we subconsciously know will never come.

Whom are we willing to sacrifice, and whom to protect, and at what price or risk?

Which of us would have helped Jesus on the day of his crucifixion? Most of us might imagine we would have joined the solemn crowd, but would our presence have been motivated by our awestruck terror at the crime of the state against the man, or would we have been drawn only by the spectacle of his crucifixion? Unfortunately, if most of us were simply spineless thrill seekers in the crowd at the foot of Calvary, social and religious pressure would discourage us from realizing it. We would sense something special in the man, but we would fear his revolutionary ideas. If we had recognized the importance of his message sooner, Jesus, rather than Barrabas, would have been set free. It is painful to consider the possibility that it is people just like ourselves who chose his crucifixion.

If too many of us fool ourselves into thinking we would not have been part of the compliant majority of the Third Reich or at the foot of Calvary, then we may be too blind to see what the controllers could do to us, once the means and the motive arise.

Freedom

Slavery Perceived As Freedom

Slavery is a matter of perception. With steel shackles it is obvious. Without shackles it may still be present. In the evolution of civilization the means of control have changed and the satisfaction of our basic needs has increased, but we may still ask if the average person has become freer with respect to the ruling class.

Mind control, through words and implicit threats, is stronger, more sure, than chains when large masses of people are enslaved. This has been known for millennia (refer back to the comments of Gheti and Terrence in the section *Witch Doctors, Kings, Priests, Presidents, and Other Liars*).

The image of enslavement through mind control is obfuscated by tolerance of a class of rebellious intellectuals. This can work well, since it encourages the image of freedom, which is so necessary to mind control enslavement. But the size of the disenchanted group cannot be too large if the society is to remain stable; indeed, its size is an inverse measure of the success of the ruling class.

Generally, societies are structured to allow most of their citizens to see things only a certain way; this is necessary for the society to exist. The stability of the society, and the success of its ruling class, is reflected in the degree of freedom its citizens feel, regardless of how restrictive are the unstated boundaries of their philosophical ideas. In our society, an eccentric point of view will be tolerated if nonthreatening numbers of people are involved. Also, if a person has too bizarre a set of ideas and is too public about it, there will be a reduced likelihood of that person getting a good job. The pressure for conformance is always with us; we habituate, so we do not notice.

As another example, if one of us knows something of which public knowledge would threaten (a member of) the ruling class, any of several things might happen: We are safe in revealing the truth if the truth or our manner is too bizarre to inspire credibility; If our knowledge is somewhat more credible, the ruling class (or a member of it) will use their superior validated position and their control of the communications media to invalidate us and our information; If (the member of) the ruling class thinks we will go public and be believed, then (he) they may attempt to silence us some other way. In most cases, one of the first two scenarios applies; occasionally the third scenario applies. A member of the ruling class may fall, but this is never allowed if this member has enough support from the rest of the ruling class. Even in ancient Egypt, pharaohs were occasionally murdered for the benefit of other members of the ruling class; such occurrences do not equate to freedom for the average citizen.

In the deep South (of the U.S.) before the emancipation of slaves, on a plantation there was argument and bickering between the slaves and the overseers. A slave might even strike an overseer. For example, if a female slave was about to be whipped for some transgression, and if she was smart and tough, she fought, scratched, and kicked the overseer all the way to the whipping post. Then she would get an extra heavy whipping, but in the future that overseer would give her more leeway, and his unpleasant experience with her would be passed on to other overseers. Some level of resistance and rebellion is found within every system of slavery.

Another obfuscating issue regarding our enslavement is the truth that any one of us might aspire to and reach a position of enormous wealth or power; any one of us can become a part of the ruling class. Certainly, average citizens have done so: billionaire and ex-presidential candidate Ross Perot is one such example. Bill Gates, head of Microsoft, is another. Jimmy Carter is a third example.

Of course, even slaves may reach high positions, whether or not they become free. For an overt example, usually the grand vizier—the chief administrative official of the Ottoman Empire—was a slave throughout his life.

Napoleon Bonaparte, a commoner, rose to become emperor of France. The English Parliament in 1656 offered to make Oliver Cromwell king.

In ancient Egypt around 1500 B.C. a commoner named Senenmut (also spelled Senmut) was the second most powerful person throughout most of Pharaoh Hatshepsut's reign of two decades. That Hatshepsut was a woman, contrary to two millennia of tradition, is also remarkable, though incidental. (She assumed the masculine royal accoutrements to perform her pharaonic functions, and thus was history's earliest cross-dresser.) Her stepson successor ,Thutmose III, was more commoner than royal, having only one quarter royal blood (Grimal; Payne). Later Queen Tiy, a commoner, exercised great influence, in part due to her marriage to the pharaoh. Subsequently, her commoner brother Ay became pharaoh, possibly through royal marriage. Later still a soldier, Horemheb, presumably a commoner, became pharaoh, and as his successor he appointed a commoner, Ramses, who became Ramses I of the Ramses dynasty (Grimal).

The potential for the political or economic ascension of any one of us does not equate to freedom for the rest of us. Our freedom or slavery has always been based on political, economic, religious, judicial, and legislative institutions, not on whether commoners or aristocrats occupy them.

The problem is this: any one of us might gain enormous wealth or political power, but not many of us practically can, since there are limited positions among the ruling class. Thus, the disparity still exists between a small number of rulers and the mass of average citizens, just as it did between the pharaoh and his (or her) citizen slaves.

Rather than the emergence of any human enlightenment or moral flowering in the last several millennia, it is just as likely that technological advances—improving means of communication, improved health, etc.—associated increases in our standard of living, and evolving institutional structures have caused a change in the visual appearance of servitude.

Well, if we are having our daily needs met, with an occasional helping of cake and ice cream, and still have time to sing and dance, is there any reason for discontent, whatever our manner of servitude? And why should we even care whether it is called slavery or freedom?

Today each person in the richer countries has more pleasures, goods, health care, and other services available to them than an Egyptian pharaoh, a Persian satrap, an Ottoman sultan, or the king of France. It is only in status that the average person still has less than such rulers; we are ruled rather than rule. The (possibly temporary) abundance obfuscates the issue of servitude, unless we were to agree that a slave is free if he is so surfeit on pleasure that he doesn't notice his shackles.

When does the contentment become so high, and the shackles become so irrelevant, that there is no further chance that the chain will again be yanked?

The average person is now trained to want and expect those things that will bind him to the will of the ruling class. We are socialized to marry, have children, and own a home. Indeed, our culture encourages us to assume (or demand) a right to these wonderful gifts, and to take them for granted. To be validated as a responsible member of society means to pursue these things, and to have life insurance and a car as good as the Joneses. These social imperatives imply financial encumbrance, which necessarily makes each of us more willing to accept the importance of the ruling class and our paid place as a cog in the machine.

All this may be well and good. It has certainly created a system (in the U.S.) which can react to global threats from similarly constructed nation-entities, and stay ahead of these entities in material benefits. It certainly allows us (in the developed countries) to take advantage of the less developed nations and the rest of Earth life. However, this system, while having provided well enough so far (ignoring at whose expense), may also blind us, as it did the

First and Second estates of France, to the question of whether it can or will continue.

What we have now may not feel like slavery. Over the centuries our labors have brought increased rewards, as expectations of us have decreased. In the next ten years the expectations will drop to zero. During this same time, for the same technological reasons, the life and death power of the ruling class over us will become greater than it has ever been. We will be totally at their mercy.

Freedoms Willingly Surrendered

By the time we have polluted, contaminated, and decimated air, water, and earth so thoroughly that we can no longer live off the land, and we then depend upon nutriments that are produced, processed, and distributed by the automated digital system run by the controllers, we will have given up most of our freedom. And we will have done so willingly, as the price we pay for having a technology based standard of living that implies such pollution.

Advancing technology will make necessary the further surrender of freedoms. As a trivial example, we can no longer drop a one pound package in the mailbox; now, we must give it to the postal clerk. Some freedoms will be slowly whittled away; others will be eliminated in large portions. All this will be done for good reason.

Today, one individual could conceivably construct and detonate an atomic bomb in New York City, or create and release a deadly virus in Los Angeles. Digital weapons have already been used by numerous people. So far, many digital attacks have been practical jokes, only of a vandalistic nature, or promoted for egotistical reasons, but too there have been successful serious attacks (refer back to *The Digital Battleground* section).

In the near future even more lethal digital and nondigital tools (weapons) will be available to average people of average means and average intelligence. The awesome power of molenotechnology and other technology advances may be viewed as very positive by some of us; on the other hand, it may be very frightening: It potentially will give Homo sapiens god-like control; however, its use may be uncontrollable. Furthermore, as we become more dependent on computers and communications, the terrorist may never have to leave his terminal to starve millions of people to death through a viral attack on computer/communications systems that will be essential to food production and distribution.

Through AI, automation, and molenotechnology, each frustrated, disgruntled, insane, or otherwise disenfranchised individual or terrorist is gaining the ability to harm a larger number of us in one keystroke. We will willingly give up freedoms in order to be protected by the rulers from the civil threat of high technology, and consequently we will become less free.

Though we will voluntarily sacrifice our freedoms in the interest of public safety, we will not necessarily succeed in establishing the hoped for security. There is no contradiction here: Our decreasing freedom may not provide enough safety to keep pace with the increasing dangers we face.

For example, despite current resistance, gun control laws will ultimately be enacted. The average person, being a good citizen, will obey. Thus, fewer impulsive people will shoot guns at anyone; however, determined people of malevolent intent will still have access to guns one way or another, and will continue to use them. And so will the police. It is only law abiding citizens who will not have them. Fewer intra-family crimes will be committed with guns.

As increasingly dangerous advanced technology capabilities arise, of course the ruling class will enact security laws whose intent will be to limit their use; this will be done by outlawing the possession and use of various capabilities without proper authorization. Good citizens will abide by such laws to be sure; however, terrorists and other determined individuals will have access to these digital and nondigital weapons of increasingly disruptive, destructive, and deadly force. (The difference and overlap between conventional, digital, and nondigital weapons is explained in the *New Battlegrounds* chapter.)

Thus, the average citizen will become more controlled, while the extremists become less controllable. On the other hand, rather than extremists, it is the controllers from whom we will have the most to fear, as we give up our freedoms and as their corporal power and control over us increase phenomenally.

Imminent Change

"The sun will rise tomorrow," is an inductive truth whose basis is simply that in recorded history it has always been so. The physical laws that have arisen to explain it have no causal effect; rather, they are only summaries of what has been. Granted, these summaries have been neatly transcribed by brilliant humans—Kepler, Newton, Einstein, Einstein-Maric, and others—into symbolic formulas. Subsequently these formulas have been given causal powers, but the belief in the causal connection is only another manifestation of man's historically proven willingness to believe in the magic of symbols.

As previously mentioned (in *The Natural Environment* section), inductive thinking is one of man's powerful survival tools; it usually works. On the other hand, induction is akin to the weatherman noticing that it is now sunny and consequently predicting that it will be sunny again tomorrow; that usually works too. Yet the sun may rise in the form of a supernova, in which case we would all be burned to a crisp.

When people are reasonably satisfied with their lives (regardless of our millennia-old complaints of taxes, conscription, roads, and the quality of our daily bread) they acquire a feeling of stability and safety, an inertia of belief about the way things are. This is both positive and negative: it engenders social stability; it also creates blindness to imminent change and danger.

Being able to see ourselves objectively is difficult, sometimes painful, but it is essential if we are to clearly view the threat the controllers pose to us. In part, this is because they are so much like us, and we like them, in what we want and how we behave.

Blindness, imposition, perpetration, and allowance are intimately connected. When one being or group seems to impose its will on another, what are the dynamics of sharing responsibility for the final result of freedom or servitude, enslavement, and death?

Usually one being asks another for something, either implicitly or explicitly, but not necessarily directly or by words. The most fundamental communication of objection is physical resistance. If there is no resistance, the imposing being proceeds to take what it wants.

But communication is equally the responsibility of the sender and the receiver; the sensitive communicator will understand even the unspoken needs of the oppressed. A being who is determined to impose its will may not be susceptible to communication of any kind; it may even claim that the victim wanted the imposition (like the rapist claiming the victim "wanted it").

A vocalized objection is usually understood, whether or not it is respected. However, it is a trait of Homo sapiens that when we are set on having something, we will ignore clear contrary communication, whether or not it is vocalized; in other words, the degree of receptivity to communication is inversely related to whether the information (or message) is what we wish to hear or believe.

The most fundamental motivation of all beings is to survive. We all know this. Yet, the average man has trained himself to ignore this communication, when he has the greater power over a being whose life he wants. Anyone would have been able to perceive that whales would rather live than be harpooned. They do not have to speak for us to know their will. But we wanted oil for our lamps. If whales had had guns, we would have been more attentive to their needs; we would have listened, or perceived, possibly without denial. We may have been angered; we may have fought; we may still have rationalized, killed, and squeezed the oil from their flesh. But we would have heard.

For our own survival it is critical to understand that man's past claim of superiority (intellectual, racial, religious, spiritual, sentient, etc.) in justifying various periods of RPM, is simply a circumlocution to avoid having to admit that we take what we want when we have the power to do so. It is ironic that our language, a significant manifestation of our proclaimed superiority, so easily provides the words to mask our moral malleability and inadequacies.

Again, this point is raised not to suggest judgement of Homo sapiens' behavior toward our fellow beings of all species, but rather to raise an alarm about the possibly false assumptions we make about whether we ourselves, the average citizens, are able to hear, or more importantly for our survival, whether we will be heard in the near future by the ruling class.

The danger we run at this point in our history is that we are only capable of seeing too little too late. With the mythical, magical, god-like capabilities that are emerging through technology, this deficiency may well prove as mortal to ourselves as it presently is to all other species.

The coming age of molenotechnology, even if controversial, will be billed by the ruling class as good for everybody. And as with many new things, all wonderful results will be predicted for it. Theoretically, it could in some sense give Homo sapiens more control—god-like control—over Earth, and that truth, together with its newness, gives it the usual magical aura and allure that has always drawn us. We are like natives in the New World, ready to bargain vast tracts of land for shiney buttons, and having no comprehension of the significance of the trade. But with all new things that give Homo sapiens increased control, while universal benefit is generally proclaimed, good or bad always depends on people deciding how the tool will be used, and by whom.

It is a pretty habit of our leaders to comment to us on the pricelessness of each human being. Why shouldn't they? Regardless of what they actually feel or believe, it feels good to hear words of our value spoken by someone powerful, wealthy, or charismatic, someone whom we admire, adore, or perhaps envy. And it is such a small thing for them to say; words cost so little.

At the same time the ruling class knows the assessed dollar value of each soldier and worker. Separating fantasy from reality is figuratively done with a butcher's knife in the private offices of government and industry. That the enthralled masses so easily accept this double-valuation system, given how our leaders treat us otherwise, is disconcerting. This is a simple example of the power of leadership, which is allowed to practice, preach, and profit from simultaneous contrary truths—such is our gullibility and the power of words.

The warm steel shackles gently being tightened about our necks are lined with silk and padded with goose down; our comfortable dreams are undisturbed, even encouraged. Within ten years we will be less free, more enslaved, more at the mercy of the ruling class than any people have ever been. As new and revolutionary technological forces come to bear, and the structure of society changes, and our importance to it declines, the danger of our present contentment may be that we remain blind until the chain suddenly snaps taut.

8

SACRIFICE

Religion, Morality, Esteem

A fundamental human characteristic is the wish for more material possessions, security, longevity, status, and so forth. Whenever we get more, we have a period of excitement, frequently accompanied by thankfulness, followed by habituation. After we have gotten used to what we have, we again want more.

Our constantly rising expectation has a way of turning into need—"We need it." Because of the emotional appeal of the word *need,* once we need something, taking it seems to become more justifiable. And in taking, we have a way of rewording so it becomes giving.

This kind of giving is seen in simian food distribution behavior in a particular species in Indonesia: the original finder of a piece of food must give it to one of the leaders, who then takes a big piece and gives a smaller piece back to the finder. This behavior, though learned, is probably supported by fundamental group survival instincts for sharing and controlling: the leader gets the pleasure of giving and being in control.

In this exchange, there is no need for the leader to announce that he is in control and has taken the larger share. Part of the leader's deal is the pleasure of telling you what he has done for you. In human society, the rulers always emphasize what they have given, so that is remembered.

A memorable example of our ability to reword what we do

appeared several years ago in one of Chevron's beautiful pro-environmental television advertisements. A deer was shown drinking from a man-made pond, and the voiceover concluded, "...people providing for nature's thirst..." The human gift of this pond was poignantly portrayed; the multitude of ponds, lakes, and rivers that have been polluted or taken was ignored. This is the idea of taking big, giving back little, and then proclaiming the total is giving, and presumably that image of humans giving to nature is the one that we will remember.

The idea that people could be the ones "providing for nature's thirst" in its very utterance presumes that all her waters belong to Homo sapiens and have already been possessed by us, the rightful and sole owners. The idea of our generously "providing for nature's thirst" is so replete with self-serving aggrandizement and blindness that, considering the relatively miniscule amount of water we have left to other species (only for the moment, of course), and the number that have died because of our behavior with water, the use of this phrase in praising human deeds is the moral equivalent of laughing with pleasure while disembowling all Earth's newborn fawns.

The ancient Greek philosopher Plato said that "what separates man from the animals is man's ability to ratiocinate." Our ability to ratiocinate does seem higher than that of most other animals; however, an equally important attribute is our superior ability to rationalize.

Humans (as individuals and societies) always find rationalizations to justify RPM. Of course, these actions are seldom described that way by the perpetrators; if RPM is recognized at all, it is relegated to an incidental and regrettable consequence of some worthwhile or justifiable endeavor. Particularly noteworthy (shocking) among humanity's rationalizations for RPM, especially as it relates to our capacity for not seeing what we don't want to see (as discussed earlier), is the frequency with which our rationalizations employ superiority of intellect, sentience, civilization, race, or religion. The implicit (possibly subconscious) arrangement is that if our proclaimed superiority is sufficient and

God is on our side, the victim deserves none of our moral behavior. This is akin to honor among thieves; it is comparable to the child's idea that it is okay to break a promise if you secretly crossed your fingers while you made it.

Thus, upon wanting something, we have proven our ability through intelligence and circumlocution to come to need it. Through religion, rationalization, and superiority we justify taking it. Then through cleverness, blindness, morality and the need for esteem, we come to believe and proclaim that the whole thing has been a beneficent act of human giving.

In our acquisitive actions we are thus supported by all that we identify as special, unique, and important about ourselves as humans. In the process of taking, we confront few moral dilemmas, and we enjoy an expanded sense of self in the success of our aggressions. Too, one of the greatest attributes and maintainers of self-esteem is to be able to give something.

Thus, God is on our side, we are successful, we take what we want, we get more, we are good, and we are givers. We should indeed feel very positive about ourselves.

It is a palpable error to believe that transgressions based on these aptitudes are part of the past, or that they lie outside of ourselves (the average citizen) and our society. Furthermore, it would be foolhardy to think that comparable transgressions against us are beyond the present or future ruling class.

Past and Present

Countless examples demonstrate humanity's past and continuing willingness, when acting individually or in groups, to sacrifice others for his own benefit whether this be amusement, convenience, control, sexual satisfaction, profit, material gain, attention, a better dinner, spiritual comfort, or to improve the balance of trade.

For those readers who might wish only to view the average citizen as fundamentally good, and who might scoff at any examples to the contrary, it is worth remembering that a majority is not needed for a repetition of any of the following examples. A minority in power, a ruling class, is sufficient to recreate any of the them. Even if each of us were nearly an archangel, the ruling class may act without our consent or with our compliance, induce us to act for them, act against us, or even induce us to act against ourselves.

These truths have been recognized for millennia. Niccolo Machiavelli (1469-1527) saw as clearly as anyone that in successful statesmanship, cunning, deceit, and carefully considered crime pay, and he knew that nations and states could be misled by crafty leaders. As in taking lambs to slaughter, it only takes a few clever plotters to lead the innocent astray; the greedy are even more ripe for destruction. He also felt "we should notice how easily men are corrupted and become wicked, although originally good and well educated." He also knew that in ascertaining the meaning of current events there is great importance in historical examples, and this connection was in large measure his motivation for writing *The Discourses*.

Thus, continuing dangers can be revealed by considering the following examples. So, despite their unpleasantness and familiarity to most readers, they are gathered below to counter our understandable tendency to keep them out of our minds. For a few moments the reader might give these examples some new consideration, even though they are uncomfortable or even unthinkable, because they may apply to our near future. (The horror of the Holocaust of the Third Reich needs no repetition here since it was covered in the *Hitler's Holocaust* section.)

(1) The Opium Wars (1838-1840)—Balance of Trade

During the period of 1838-1840, the opium problem in China came to the boiling point as China attempted without success to enlist the support of the Western nations to cease their importation of opium into China. The conflict resulted in the Treaty of Nanking (1842), which gave Western nations the control they needed to guarantee a continuation of trade with China on terms made favorable for the continuing importation of opium.

Some of the ugly truths leading up to that treaty are as follows.

> British-run East India Company took control of the the drug's production in India.... By capitalizing upon a massive addiction to smoked opium in China—and in substantial measure helping to create it—England and the other Western nations shifted the balance of trade in their favor (Wilson).

Lin Zexu (Tse-hsu), China's commissioner for foreign trade, appealed to Queen Victoria in a letter:

> Let us suppose that foreigners came from another country, and brought opium into England, and seduced the people of your country to smoke it. Would not you, the sovereign of the said country, look upon such a procedure with anger, and in your just indignation endeavor to get rid of it? Now we have always heard that Your Highness possesses a most kind and benevolent heart. Surely then you are incapable of doing or causing to be done unto another that which you should not wish another to do unto you (Chang Hsi-pao quoting Lin Zexu).

Lin was never given an official reply. When he subsequently caused 20,000 chests of opium to be destroyed, England sent her mighty ships-of-the-line and beat China into submission.

(2) Cigarettes—a Century of Profiting from Death

The invention of the cigarette rolling machine in 1872 inspired resourceful marketing people who sought to sell more cigarettes. Their vision had both positive and negative effects: The increased use of cigarettes created jobs and it killed people.

At that time the health hazards were perhaps less familiar, but they were certainly not unknown. The chemicals in cigarette smoke had already been analyzed, yielding a nasty list. In the December 1871 *Scientific American* appeared the following comment:

> In view of the enormous and increasing consumption of tobacco it has become a question of very great importance what effect upon the general standard of health is produced by it... and though our confirmed taste for smoking and our natural desire to find it a harmless practice have led us to peruse with particular care all that has been said in its favor, we avow that

> neither reading nor experience has convinced us that the general use of tobacco is other than an unmitigated evil.

In the 1880s The American Tobacco Company began putting baseball players' photographs in cigarette packs. An incidental effect of this marketing zeal is the rare 1910 Honus Wagner baseball card. Since the cards were created and used without Wagner's permission, and since he was very much against smoking, he insisted on a recall of the cards. Thus, only twelve of them exist. In 1991 one was auctioned for $451,000.

A century after the awareness demonstrated in *Scientific American*, an apparently relatively impotent arm of our government began to resist the tobacco industry to little avail:

> A third of a century has passed since the first U.S. Surgeon General's report on smoking persuasively assembled the scientific case on the lethal effects of the habit. Yet the rest of the Federal Government, deftly manipulated by the powerful tobacco industry and fearful of antagonizing the industry's tens of millions of addicted customers, has allowed the cigarette to remain our most deadly...product (Kluger).

The tobacco industry through the years had prevaricated, dissembled, and lied.

> Its executives, in what amounted to a premeditated conspiracy to disinform the American people, continued to deny what they and their scientists...knew to be true about the addictive and fatal nature of their product (Kluger).

Through the decades many public and health minded organizations challenged the tobacco industry's ghastly existence.

> But it wasn't until the spring of 1994 that the world got solid proof of tobacco-company wrongdoing (when several key people each) received an anonymous package...some 4000 pages of secret documents from Brown & Williamson Tobacco Corporation.... The papers, dating from the 1950s, showed that the industry's own researchers knew about the addictive qualities and health risks of tobacco before independent researchers could confirm this information (Bartholomew).

To executives willing to lie about the addictive power of nicotine and the lethal effects of smoking, the idea of increasing the nicotine content must have seemed only natural—a good business plan:

> Whistle blower Jeffrey Wigand, the former Brown & Williamson executive...wielded company documents to help press his claim that the tobacco company had deliberately manipulated nicotine levels (Smolowe, 27).

One might take the point of view that we are all thinking adults, able to make wise decisions even through a boldface deluge of deception. On the other hand, countless studies, common sense, and the very existence of the advertising industry prove that we are not so capable. When we are lied to or denied information, our freedom of choice is lessened. Even less capable are our children, whose bodies the industry also wanted:

> ...the obvious but startling admission by one of the smallest tobacco companies, Liggett Group, that cigarettes are addictive and have been pointedly marketed at kids for years (Smolowe, 28).

The cigarette industry is an obvious current example of what those who are in control will do: They have repeatedly shown a willingness to kill any of us for a profit. Over a ten to twenty year period they have and will allegorically cut out our lungs and eat them, all the while ghoulishly smiling and denying there is any blood dripping from their chins.

Equally horrifying is the proven willingness of the government to support the tobacco industry's death machine. Was it only at long last the acknowledged red ink of Medicaid expenses for smoking-related illnesses that raised the protective shield of government? Ferguson, an avowed smoker himself, takes a harsher view:

> I see a carnival of dissembling and bad faith...the state attorneys general, who brought the suits supposedly to win back the funds drained from state treasuries by smokers hacking away their Medicare dollars...are by and large second-tier pols (politicians) on the make, grasping for the kind of publicity that

> might boost them to the Governor's mansion or a Senate seat, but they know as well as the nearest actuary that smokers save the treasury money by dying young (Ferguson).

This latter suggestion is startling, but it is easy to understand. Medical expenses tend to mount for the average aging person, whatever the final cause of death. Medicare pays only in those cases where the person is old enough. Thus, not only do we realize that money given to politicians and parties over the years encouraged lawmakers to leave us vulnerable to tobacco industry death camp mentality, but too the government budget profited by our earlier deaths. We must wonder if that is one reason why this profitable industry will be allowed to continue under the recent government-tobacco industry proposal described by Kluger. Perhaps an additional reason is that the export of death benefits our balance of trade:

> American cigarette sales internationally are rising 3% to 5% a year, and companies such as RJR are grabbing half their revenues there already (Kadlec).

As intimated in the Blindness chapter, the ruling class cannot be held solely responsible for the cigarette industry's macabre travesty of commercial enterprise. A large fraction of Homo sapiens' believes, even today, despite all the medical facts, that they can smoke with impunity.

But smoking is only one example of a common human attribute (as mentioned earlier). We all, to a larger or smaller degree, play some kind of time lag dance with death. Wherever there is a lag between promised benefits now and deadly cost later, many of us will behave as though the later deadly result does not exist. The lag time between denials, truth, and reality will continue to allow windows of opportunity that will be exploited by the ruling class, and we will let them. (The biggest time lag exploitation ever imagined will be discussed in the *Incidental Holocaust* section.)

Humanity is gullible and greedy, and strongly affected by pretty words and pictures, which the ruling class is very willing to offer.

If we are told and shown often enough that smoking is romantic or heroic, and if the true macabre price is distant, many of us can be convinced that the bill will never come, or that it will never have to be paid, even though this belief has been proven scientifically false.

(3) Munchausen Syndrome by Proxy (MBP)—the Need for Attention

We should not be surprised by the true horror (or doubt) of what we might yet do to others, or what might yet be done to us. Even mothers do it to their own children, when it satisfies the mother's need for attention. Munchausen by proxy (MBP) disorder is described in a shocking *U.S. News & World Report* article:

> In its mildest form, Munchausen by proxy (MPB) may lead a parent, nearly always the mother, to invent symptoms for her child, forcing doctors to perform unnecessary tests or to prescribe unneeded drugs. At its worst, MBP compels a parent actually to induce illness, sometimes with appalling cruelty. Mothers have smothered and then revived their children, injected them with bacteria, or induced diarrhea, seizures and vomiting with a variety of over-the-counter drugs. One child was treated for five years for inexplicable sores before it was found his mother had been applying oven cleaner to his back.... But what is certain is that people with the disorder have a powerful need for attention, and being a brave mother of a sick child provides that.... Even when caught in the act of harming their children, most mothers tearfully deny any wrongdoing (Brownlee).

Giving the condition associated with this parental behavior a name, and calling it a disorder, is very useful to its identification and clinical discussion, and to its possible cure. On the other hand, there is a danger that we depend so much on the implied causal connection between the label and the behavior, that we fail to realize that, even where the disorder has not been identified, a threat exists.

(4) Child Abuse—Sexual Satisfaction and Control

Comparable ugliness exists in sexual molestation and torture that modern society now recognizes many parents do to their children (Bass & Davis).

(5) Eat the Dogs—for Glory

Admiral Richard E. Byrd (1888-1957), the renowned Antarctic explorer, modeled himself after Roald Amundsen (1872-1928), his Norwegian polar predecessor, who "conquered" the south pole (Rodgers). Amundsen's standard practice was to use his sled dogs for food. For the efficiency of operations, dogs were used to pull sleds and provisions, and then as the expedition's sleds grew lighter and fewer dogs were needed, they were included in the meal plans; recipes aided the appetite. Dogs were expendable pack animals.

Dog slaughter planning was fundamental to Byrd. After dogs served the expedition purposes as living creatures, they either became food or simply became too expensive to feed. Rodgers, in his book *Beyond the Barrier*, about Byrd's first expedition (1928-1929), writes of the grief some men felt in slaughtering animals who had served loyally beside them through perilous Antarctic journeys for the glory of their masters; nonetheless, despite the tears, such slaughters were commanded, expected, and executed (Rodgers).

The ship *Bear* arrived in Antarctica for the evacuation at the end of Byrd's 1940-1941 third expedition. There were again 'too many dogs' to keep. Some of the dogs were taken aboard. The excess were put in harnesses on the ice and a timed explosive charge was set to blow them up after the ship departed. (Bertrand)

Dog meat was not unique to the Antarctic. Commander Robert E. Peary (1856-1920) also planned his Arctic polar sled trips carefully. To maximize his probability for being the first to reach the North Pole and return safely, he had to minimize expense and time on the trail. He ate some of his dogs on the return trip during his 1893-1895 polar expedition (Brearley), whether this was by premeditation, negligent planning, or recklessly risking guileless lives.

Most dogs have a loyalty and willingness to work that surpasses man at his best. This particular trip symbolizes man at his worst.

(6) Bad Wolves—they must be Bad, they want to Eat

Homo sapiens has maligned wolves for millennia. For example, ranchers in the continental United States get very self-righteous about wolves feeding on their livestock. These ranchers feel very protective of their interests, and have loaded weapons and immediate impulses to kill the wolves. This is understandable if one's focus is on present legal property rights (established by Homo sapiens law—uhmm, no, established by WASP law subsequent to stealing the land from Native Americans), but let's step back for a moment. Before our ancestors (or Native Americans) came to this land, wolves were able to hunt wild game on the same land that these ranchers now so righteously claim is theirs, and onto which the wolf is accused of trespassing. At what point did the wolf become a problem? From the wolf's point of view, when did man become a problem? Who was willing to share the land, even if through ignorance of things to come? Who thought nothing of taking everything that was there, and to justify RPM with fine ideals, emotional appeals, clever words, religion, and self-righteous indignation? (These things all depend upon our intelligence.)

(7) Who is Eating all the Fish?

In 1978 the *San Diego Tribune* carried a story entitled, "Fishermen collect $12 bounty for each dolphin in slaughter." Japanese fishermen drove a thousand dolphins onto an island shore, and systematically stabbed and clubbed them to death. The reason given for this atrocity was that the dolphins were eating all the fish! This is the age-old problem of maligning any other beings (including wolves) who might want the same thing we humans (or our particular survival group of humans) do. Now, who was eating all the fish?

More recently in the United States in Monterey County California, fishermen had a similar vicious response to pelicans and sea lions, whom humans blame "for eating all the fish."

(8) Chief Seattle—Whitewash in Red

White man's invasion of the American continent, subsequent conquest of Native Americans, and annihilation of many of them and their way of life, was carried out with a feeling of self-righteousness and moral purpose. Our frequent references to religion, and to the use of the word "heathens" to describe the Native Americans, were not verbal contradictions to the Bible, and thus they served as brilliant circumlocutions that helped us feel that God approved of what we were doing, despite the obvious fact that the Bible offers no support whatever for RPM against those of another culture; indeed, the New Testament stands directly again RPM in any shape or form.

Chief Seattle's (1786-1866) speech to Governor Stevens (given in *The Enlightened Mind*, edited by Michell) conveys a concise revelation of the truth, lie, and persecution in White Anglo Saxon Protestants' RPM of Native Americans. This man of primitive origin, heritage, and existence saw as clearly as any modern man—politician, entrepreneur, clergy, or layman—ever will about how it all works, how it all fits, how it all lies.

> ...The White Chief's son says that his father sends us words of friendship and goodwill. This is kind of him, since we know he has little need of our friendship in return. His people are many, like the grass that covers the plains. My people are few, like the trees scattered by the storms on the grasslands.
>
> The great—and good, I believe—White chief sends us word that he wants to buy our land. But he will reserve us enough so that we can live comfortably. This seems generous, since the red man no longer has rights he need respect....
>
> But I will not mourn the passing of my people. Nor do I blame our white brothers for causing it. We too were perhaps partly to blame. When our young men grow angry at some

> wrong, real or imagined, they make their faces ugly with black paint. Then their hearts are ugly and black. They are hard and their cruelty knows no limits. And our old men cannot restrain them.
>
> ...Young men view revenge as gain, even when they lose their own lives. But the old men who stay behind in time of war, mothers with sons to lose—they know better.
>
> Our great father Washington—for he must be our father now as well as yours, since King George has moved his boundary northward—our great and good father sends us word by his son, who is surely a great chief among his people, that he will protect us if we do what he wants. His brave soldiers will be a strong wall for my people, and his great warships will fill our harbors. Then our ancient enemies to the north—the Haidas and Tsimshians—will no longer frighten our women and old men. Then he will be our father and we will be his children.
>
> But how can that ever be? Your God loves your people and hates mine. He puts his strong arm around the white man and leads him by the hand, as a father leads his little boy. He has abandoned his red children. He makes your people stronger every day. Soon they will flood all the land. But my people are an ebb tide, we will never return. No, the white man's God cannot love his red children or he would protect them. Now we are orphans. There is no one to help us.
>
> So how can we be brothers? How can your father be our father, and make us prosper and send us dreams of future greatness? Your God is prejudiced. He came to the white man. We never saw him, never even heard his voice....
>
> But why should I mourn the passing of my people? Tribes are made of men, nothing more. Men come and go, like the waves of the sea. A tear, a prayer to the Great Spirit, a dirge, and they are gone from our longing eyes forever. Even the white man, whose God walked and talked with him as friend with friend, cannot be exempt from the common destiny.
>
> We may be brothers after all. We shall see (Mitchell, editor, adapted by William Arrowsmith).

There were no war crimes trials to prosecute the white man, because the white man won. The winner always views itself as having done the right thing, and writes the history books to reflect this view.

(9) The Simulated Prison Study—Enjoying Power

At Stanford University in Palo Alto, California in the early 1970s, a psychological study was conducted that implied some very negative things about average human characteristics. Professor P.G. Zimbardo's study was nominally about prisons, but neither real prisoners nor real guards were involved. Instead, twenty-four normal male college students were screened from an applicant pool of seventy-five for the study in which an artificial prison was created in the basement of the Psychology Department on campus. The roles of guards and prisoners were randomly assigned. Nonetheless, the guards quickly lost respect for those over whom they had control; the guards became abusive and enjoyed tormenting the prisoners. This study's implications about the potential for abuse by an average person is rather unsettling, but cannot be ignored simply because of the accompanying feeling of discomfort such an idea creates (Haney; Banks; Zimbardo).

> Zimbardo suggested that the reason for the deterioration in guard behavior was power. The guards were able to exert control over the lives of other human beings and they did not have to justify their displays of power.... After day one, all prisoner rights became redefined as privileges, and all privileges were cancelled (Banyard; Grayson).

(10) Spiritual Satisfaction through Human Sacrifice

The shiney black obsidian knife blade hesitates high against the bright blue sky. Surrounding the base of the pyramid, the awed religious crowd holds its breath. The knife plunges sharply down into the flesh of the 12-year old virgin. The skillful holy man carves adroitly, and lifts the girl's still beating bleeding heart into the air for all the congregation to see and marvel.

There is no clear evidence for a moral instinct; at least, there is none that cannot equally well be explained in terms of long- or short-term self-interest, either for the individual or the society. For anyone who has not been trained to feel horror in the spectacle of human sacrifice, the instinctual intelligent animal response to it must be one of pleasure. This response would be quite reasonable since the best way to gain new appreciation for something we have and take for granted is to lose it, and then to get it back; having somebody else lose it is a good substitute for promoting this appreciation. (Whether they get it back is not so relevant to us.) The unspoken subconscious relief and release is, "Boy, I'm glad it wasn't me."

The pleasure of being alive is certainly reinforced by the proximity and witness of death. This is part of what drew people to the spectacle of public hangings. This is probably as much a motivation of man's lust for hunting as ever was his need for provision; besides, who would dare risk questioning such leadership? A generally problematic truth of Homo sapiens is its feeling of empowerment in witnessing those who are less fortunate. One might speculate that this is the driving instinct behind everyone's quest for status or even their association with charities.

(11) The Joy of Slaughter

The death-pleasure suggestions of the preceding paragraphs are powerfully confirmed by a stunning article, "A Matter of Life and Death." The author, Ross Herbertson, is an avowed slaughterer and hunter, one who is very serious about his killing and who obviously finds pleasure in it (though presumably he would deny this), one who gets a spiritual thrill from "the power transpiring in that moment," and one who through his attitude and poetry has converted (or elevated) killing to a religious experience. In a soothing bedtime story voice, to his intended victims he offers slaughter prayers.

Herbertson holds the progressive and commendable position that no form of life is superior to any other. He states that all life is one, every living entity is sacred, and for his intended victim (calf or cabbage) any day "is a good day to die." He seeks and speaks to his victims, "not as a master who comes to take your life, but as a brother who seeks to join our lives together."

Perhaps it is fortunate that in our civilization he is welcome to feel as he does and is allowed a forum in which to write about it. He deserves the highest praise for so honestly and beautifully revealing what many persons probably feel, but are reluctant to admit.

Of a lamb whose throat he cut, neck broke, and spinal cord severed in the midst of slaughter and up to his elbows in pulsing warm blood, he writes,

> I knew in the depths of my being that I had just killed my newborn child. The blood that I spilled was not like his, it was his. I had murdered my child. All life truly is one (Herbertson).

In his thoughts are phrases such as, "the reverent process of this life-defining event," which accompanies his slaughtering an animal. More poignantly he writes, "One is never more exquisitely aware of life than when in the immediate presence of death." The substantial truth of this statement is one of the likely emotional bases for the evolution of sacrifice (human and other) as a religious ritual in various societies.

The potential for profound spirituality in any human activity, good or bad, cannot be denied. Herbertson has indeed found and enjoys his spiritual roots in his killing, which he refers to also as "murder." His perceptive truths and deeply religious viewpoint are clearly and poetically expressed. For him the pleasurable act of killing admittedly ranks in importance with sex and wealth.

His slight insecurity about his beliefs only reveals itself in an oversight in one paragraph, in which he effectively states that the death of one living entity (plants, animals, people) is no different than that of another. He proceeds to claim that you either do kill or you don't, and adds that even someone whose inclination is to not kill, still inadvertently kills millions of organisms in the daily process of living. However many it is, each individual human inadvertently does kill many organisms daily and, as an aggregate species, we exterminate hundreds of species daily. For Herbertson there is no distinction between killing more or killing less; that is, he denies any value to killing less. Consequently, it seems that for him there would be no distinction between killing something inadvertently and killing everything intentionally, so long as we "immerse ourselves in conscious gratitude for this incessant passing of the life force through us."

This potent article reveals a seemingly deeply religious and spiritual man of vision and courage, and a gentle man. Its profundity forces further consideration of several questions about his and our fundamental nature. With our avowed powerful use of words we might suggest any truth about ourselves and our goodness. On the other hand, when it comes time to truly comprehend the universe in all its magical workings, with our words and control cast aside, we must ask ourselves what we would feel if ever we were to exchange places with doomed creatures about whom we mumble so reverently.

Herbertson is one of those people whose behavior is so beautifully and assuredly self-expressed in his poetic avowals of equality, oneness, and brotherhood with his victims that he will be unable to sense the superficiality of his articulated spirituality, or the meanness of his behavior, until someone stands in front of him or his family with a loaded gun and the intent to kill saying, "Today is a good day to die."

Mastery, control, brotherhood, and joining lives together—these are powerful emotional concepts. Denial of the first two and affirmation of the latter two create a wonderful spiritual impact and is bound to have a powerful effect on both cognizant parties to a killer-victim relationship, but those effects will be very different—certainly not equality.

A further perspective might be acquired by hypothetically considering the potential therapeutic value of Herbertson's philosophy if applied to serial killers. Suppose, for example, such religious and spiritual thinking were offered to Jack-the-Ripper and others. Some of them are thought to feel extreme guilt for their crimes, doubtless because of society's proclaimed attitude about murder. One must ask if Herbertson's philosophy could bring peace to their troubled souls. Subsequently, attention to their emotional problems might be more fruitful.

Also, for comparison with Herbertson's avowed beliefs, it is worth mentioning an important example of spirituality and sensitivity within the Third Reich. Karl Wolff described Himmler's witnessing the execution of Jews by an Einsatzkommandos killing unit (prior to the use of gas chambers). "Wolff watched Himmler jerk convulsively and pass his hand across his face and stagger. He went to him and drew him away from the edge (of the grave)" (Padfield). Symbolic of Himmler's proximity to this "life-defining event" is the bit of brain Wolff reported to have wiped from Himmler's cheek. His empathetic response to "the power transpiring in that moment" is undeniable, as it probably would have been for any of us.

Herbertson claims that his killing is part of the reason he is such a gentle person. Doubtless, he is a fine person and comfortable to be around. One would probably find him kindly and loving toward family, neighbors, and friends. For comparison again, it is likely that most of us would have found the same attributes in Himmler if we knew him socially rather than as one of his victims.

Herbertson's existence and confessions prove that humans can at once be caring, articulate, intellectually potent, deeply spiritual, reverent, artistically creative, and poetic, and still enjoy their control and mastery over their victims and the thrill of their slaughter.

Implications for the Future

These above examples typify what we as humans have done and are doing. There is no value in using them for judgement against Homo sapiens; rather, their importance lies in what they can tell us about ourselves, which if we fail to objectively view will leave us vulnerable to humans doing the equivalent of these things to us in the near future.

Herbertson serves as an example and powerful reminder that there are people among populace, religious leadership, politicians, and the wealthy who will approach us with reverent manner, spiritual conviction, sincerity, and honeyed words, and who are intent on taking from us our assets, our bodies, our souls, and our lives.

The ruling class (the controllers) will take advantage of the rest of living beings (us or others) to whatever extent they are able and allowed. The average citizen is either being taken advantage of, or is being led to take advantage of others. In both cases, through laziness or disinterest, our unwillingness to question is always at issue, and thus our participation is voluntary. If we wish to continue denying our own willingness to sacrifice others, then we must make a larger effort to change our behavior to avoid the frightening pitfall of our denial. The most magnanimous interpretation is that we are a compliant majority, but we are still serving a slaughter house into which we ourselves will soon be thrust.

It is not sufficient to say, "But, no. That was then. That was them. I wouldn't do that. That's not me (us), I'm (we're) not like that. We're not like that now." It is always easier to see the failings or cruelty of others, even when they are separated from us in time or space, than to identify our own. These perpetrators did (do) not think they were doing anything wrong. They did (do) not see themselves as "being like that." (The reader may wish to reread Himmler's Posen Address in the section *Hitler's Holocaust.*) Of course, our reasons for doing what we do seem more justifiable to us than the reasons used by the Third Reich, for example; that is because they are *our* reasons; thus, it is natural that they feel more compelling. It is difficult to view yourself objectively when you are caught up in ongoing perpetration that your superiors, peers, society, and religion approve of or insist on.

Unfortunately, ugly aberrations, in analogy with MBP, do not exist only when we see them. They are with us always, surrounding us now. The most frightening proposal is that such behavior is not aberrant; that it is normal and happening right now, in ways that we do not choose to recognize. Perpetrators *in situ* usually do not recognize it.

In analogy with MBP, political leaders "have a powerful need for attention." They also almost always "deny any wrongdoing…

even when caught in the act." A leader may tearfully proclaim to heaven to cherish us, even while acting contrary to our best interests. Perhaps any of us lacking the proper balance of forces could behave this way if our needs, for attention or whatever, become great enough.

Some people may argue that these ideas and preceding examples are ugly, but that they pale next to the wondrous advances of human civilization. Whether or not that thought has any validity is irrelevant to the main point: Such things can happen again, and because of technological advances there is an increasing chance for them to happen suddenly, globally, before we become aware enough to prevent them.

Encouraging our freedom to separate ourselves from these things, to give ourselves reassurance that we are not like that, is that we may now feel shame, guilt, or regret for what we (or our ancestors) have done. For example, it is the vogue now to be sorry for what our ancestors did to the Native Americans. But it is easy to be sorry later, after the dirty work is done and we are enjoying the benefits, which we have no intention of relinquishing. As a token remembrance we created the Indian head penny and put the image of a buffalo on a nickel with an Indian head. Use of these coins shows how endearing and honorable our behavior can be, after we have taken what we want and the barrier to our wish fulfillment has been annihilated. The state animal of California is the grizzly bear, whose image is on the state flag; the last grizzly in California was shot in 1922. Today the U.S. Postal Service decorates stamps with species doomed to quiet death; for the survivors there will be zoos and reservations, just as there were for the Native Americans.

When a Nazi war criminal is caught, tried, and punished we feel some measure of satisfaction. There is a danger that this satisfaction is due to feelings of vengeance, circumlocutory self-serving implications that we ourselves are not like them, and the consequent reassurance that we are stamping out the last traces of evil and serving a punitive example so that such heinous crimes will not happen again. Toward this latter goal, our energy might be better spent attempting to open our own souls, examining our own hearts, and discovering our spirits.

Furthermore, punitive examples have little deterrent effect on members of the ruling class, who feel they are only answerable to themselves, that they are the law, or that they are above it. To fully understand this attitude we only need consider how relatively minute is our own concern for all humans who are disadvantaged with respect to us, and how our Christian religion and human laws place us above all nonhumans. An additional well-known criminal problem, especially as it relates to deterrence and recidivism, is that the criminal is convinced that he never will be caught.

Certainly, we want to believe that no one, least of all, powerful charismatic leaders who make us feel all gushy when they walk into the room, would do those things to us. But we must consider the possibility that Nazi war crimes, rather than stemming from aberrant personalities, were instead aberrant manifestations of normal personalities, like ours, or like our leaders' (or the controllers'). This idea is contrary to Miller's analysis and is too repulsive for most people to consider for more than a moment without discounting it. The possible validity of such a speculation is made less relevant (and thus less threatening) by the more plausible idea that it is nearly exclusively aberrant personalities who are drawn to positions of power; of course, neither is that a comfortable thought.

Even if Miller's impressive analysis were accepted, that the perpetrators of the holocaust were aberrant personalities, that conclusion does not preclude the possibility of normal people (specifically of the current ruling class or the controllers) from exhibiting similar aberrant behavior. In any case, the question still becomes whether, when, and how the ruling class's aberrations will again manifest themselves, and what motivation and technological power will be sufficient for this manifestation to occur.

Miller's analysis seems solid as far as it goes, but her conclusions fall short of an important interpretation: (1) that while two different people may react differently to the same circumstances, it may be their differing perceptions of threat in common circumstances that is the fundamental issue; (2) it is one's perception of threat to survival that is affected by childhood and other experiences.

Intel chieftain Andrew Grove was referring to himself, and others who rule and run our technologically advancing society, when he said "only the paranoid survive" (Gilder). While the context of this quote might be corporate competition, the paradigm is set, and it is easily applied to other facets of a person's existence. All the more reason to think so is Gilder's claim that for Grove this belief was "born in the vortex of the Hungarian Revolution."

The perceived severity of the threat to one's survival is strongly correlated with one's warped psyche when aberrant (criminal) behavior is involved. Indeed, a fundamental distinction between the healthy and insane is the survival issue; the warped psyche just feels that survival is the issue more readily than a healthy person in a given situation. Once a person's survival button has been pushed strongly enough, that is, in the sense of the degree of threat that person feels (albeit this would be under two very different circumstances for the healthy and the insane), there may be only a small difference in the resulting reaction.

It may well be that any of us would respond equally radically to a perceived equal threat to our (self's, family's, or nation's) survival. Certainly, in times of war, when the idea of such a threat is promulgated by the ruling class, recognized authorities, and church, and murder is condoned, and it becomes illegal to dodge it, the average man has frequently gone out to murder "the enemy"; indeed, we are actually decorated for such "heroic" acts. Moral precepts do not inhibit such murder, even among nations that claim to subscribe to a theology of peace and the Golden Rule. Of course, threats to our survival are often perceived in terms of someone (or entity) who (that) has or stands in the way of enhancements to our survival.

So, in the evolving technological environment, the question

again arises as to the circumstances under which the ruling class will feel such a threat to themselves, and from whom. Certainly, technology and the global situation are evolving so radically as to produce previously unforeseen threats for all of us.

It takes willingness and energy to consider these disturbing possibilities. For some of these ideas and for many of us, the demand is too great: We wish to be left alone. It is easier, feels safer, and is certainly more comfortable at the moment to have our heads in the sand.

Whether the problem is one of aberrant personalities, aberrant behavior of normal personalities, or the ugliness is normal and not always recognizable, and whether it is all of us, only in the ruling class, or in just a small fraction of the ruling class, the conclusion is the same: There is potential great danger for the large majority of us in the unstoppable near-future advances in technology.

9

GLOBAL HOLOCAUST

Three words we will soon know well: holocaust, genocide, and pogrom.

Holocaust means "a thorough destruction involving extensive loss of life."

Genocide means "the deliberate and systematic destruction of a racial, political, or cultural group."

Pogrom means "an organized massacre of helpless people."

The present ruling class, the new emerging ruling class, and every other definable human group is a potential predator or victim. Any coherent group can sense vulnerability in a potential victim as surely as does any predator with or without cognitive power. Whether this is instinct or learning, whether it is subconscious or articulated, and whether we act on our awareness or not, the simple criteria are these: (1) The victim group must not be needed as an ally to aid the predator group in its survival; (2) The predator group must have and be able to retain sufficient power over such a victim group; (3) The victim group must be one whose members are identifiable, and could be labeled as unique and separate enough that their victimization will neither alarm other similar potential victims or allies, nor cause unwanted fears in any of the compliant beneficiaries.

Society in Ten Years

This section summarizes much of what has been discussed in previous chapters.

Within the next ten years, advances in technology, largely in molenotechnology, AI, automation, and global computer processing/communications will place god-like powers of creation and destruction in the hands of Homo sapiens.

The average person will dream of having this power, feeling the pleasure of it, and will be promised it as an inducement along the path to its development; however, unbridled access to advanced technology cannot be allowed. Advanced technology will give Homo sapiens incredible control, but its use will be difficult to control. The dangers will become painfully apparent through catastrophic accidents and terrorist examples. As technology continues to develop, it will become clear that it is unsafe to permit us either access to all the tools or technical knowledge of them.

Restrictions will be imposed by law, for which we will vote in the interest of mutual assurance and public safety, and by secrecy. The use of secrecy will be very important to the survival of the cabals, who will want to hide many capabilities from one another and from the governments, whom they will quickly come to see as competitors.

Thus, access to many of these magical powers will be limited to an elite group of immortal people—the controllers—and they will expect us to trust them.

Our increasingly digital existence implies further limitations in our freedom. The privacy of communication through any medium will be dubious since it will all be routed, tracked, sifted, and stored by the controllers; since use of encryption for privacy will be restricted, we will never know what is private and what is not. Furthermore, it will be increasingly easy for the controllers to selectively disrupt our communications. These issues become much more relevant as our common interest groups become less local, more global, and we consequently rely increasingly on digital communication for our security and alliances. Also, the production and distribution of our nutriments, and other consumables, will be digitally controlled and subject to impromptu interruption.

Thus, within ten years, as an incidental but important result of the evolution of technology and society, we will be less free, more enslaved, more at the mercy of the ruling class, than any people have ever been. The controllers will gain the power of immediate life and death over us.

Some of the fruits of the technology may be ours; however, in the evolution of technology there is always a dichotomy between technological capability and economic feasibility. Miraculous capabilities will exist and magical products will be produced, but at least initially most of these will not be affordable to us, and many of them will never be allowable to us.

Homo sapiens will eventually have relatively unlimited production capability, but this will follow the complete destruction of the living planet. Technology and production will proceed rapidly enough for toxic waste to kill the planet, but not before the technology exists for Homo sapiens to survive in artificial environments. However, by the time artificial environments are needed for survival, the limited survival-environment production capacity will probably serve only the controllers.

Technology will become so sophisticated that civilization and its progress will not need us. We will have neither the aptitude nor the motivation to serve any useful function in what the controllers will define as the advance of civilization. We will then be expecting goods and services without being able to offer any useful return. Regardless of the politicians' proclamation of the value of each human being, to the controllers we will be expendable; indeed, we will be a detriment.

Morality of the Controllers

In the near future, through various complex changes in our technological civilization, most of which will be the result of things that we ourselves will demand, most of us will automatically become members of a group that the new ruling class can feel safe in victimizing to the limit of anyone's imagination. There is every reason to expect that they will take maximum advantage of their position.

Herds, packs, groups, clans, tribes, and societies formed out of an instinct for survival. For Homo sapiens over the millennia, this propensity expanded to encompass a desire for enhancements, including increased safety, health, comfort, and luxury. Consequently, the size of societies grew.

To a great extent, the ideal size of a society is interrelated with its environment and the level of its technology. In the growth of population within any given society there is a balancing between many fundamental pressures: a general desire to have children obviously contributes to growth; birthing contributes to growth; awareness of hygiene allows growth; food sources (more basically, arable land) allows growth up to some supportable limit, as do other resources, such as water and minerals (once technology has advanced that far); advances in medicine encourage growth up to the limit of the society's ability to fulfill medical demands; technology at any given level of advancement has its ideal size of labor force to support, and yet not to impede, its progress, and this plays at least an indirect and probably weak role in determining a society's population; and the presence of large threatening neighboring societies encourages growth to satisfy military demands. These issues all play a role too in the limits of how large a society can be, or is willing to be, if it is to be governable, that is, without rebellion, secession, or other dissolution.

Morality, that is, a set of moral precepts, is an important ingredient to the glue that holds any society together. It nominally dictates the ideal behavior of the members of the society toward one another to maximize their mutual safety and benefit. It also helps maintain adherence to the society's internal goals and sustain the society's competition with other societies. It may be incidental that in association with the safety we find within our group, and the suspicion we feel toward other groups, we often lack regard for and have a willingness to exploit other groups.

Morality is a somewhat fluid substance, as it must be to fill its role, gelling and liquifying according to the material or spiritual need of the times. Also, the boundaries of the group, the limits of our brotherhood or society within which a given set of moral precepts is presumed, may shift suddenly to support our enhanced welfare. Thus, ubiquitous moral potency does not exist, since morality always depends on the changing breadth of one's brotherhood, and history has shown that those who lie outside may expect no mercy.

By analogy, even in the behavior of a single person, we can see the fluctuating application of behavior codes. Suppose you are driving your car down the highway. Someone carelessly cuts in front of you, causing you to swerve. You swear, speed up, beep your horn to get his attention, and shake your fist. When he turns his head toward you, you see it is the person who always sits next to you and your spouse in church. Why do you feel embarrassed now? That is the difference.

Many of us feel no innate need to behave respectfully, or responsibly, toward someone outside our group; rather, respect toward others is learned. Otherwise, if they are not familiar to us, if they are different from us, if they are not near us, then they are from a different group. If they are from a different enough group, we need not respect them; we may take whatever we have the power to take from them. Our group-specific morality no longer applies. And that is how RPM is possible, even for people whose avowed golden rule is to "do unto others as you would have them do onto you."

Even within a society there are subsocieties, organizations, groups, etc., who function for their own benefit. Sometimes they live within the restrictions of the laws of the larger society. On the other hand, frequently, having balanced the potential penalties for breaking the laws, the risk of being embarrassed, arrested, or convicted, and the potential gain, they act to maximize their benefit. In fact, institutions even publicly do this; they simply treat the penalty for breaking a contract, for example, as part of the contract, which it sometimes is.

Globally, across our nation-societies, the ruling class has always viewed itself as a separate group, apart from us. Even in times of conflict between nations, the ruling class has had its own bond, its own special place and treatment. This has historically been so, and it is true today.

The ruling class will constantly say to us that we are one people, one group, as long as it is in their interest to make us think they believe so. They know the danger and suspicion that comes with being from different groups. They know what we do to groups that are not like us. They know we know RPM always lurks between disparate groups.

The ruling class is just like us in their propensity to take advantage, but they are superior in their skill for doing it. They will take whatever they can from others. With this goal in mind, they will continue, implicitly or explicitly, to form strategic alliances among themselves. These alliances will pursue their maximum benefit, a pursuit which if successful is at the expense of other societies and groups. It is the ones that do this well that survive. By definition, that is the structure of the world.

The momentum for morality and ethics, beyond what was mentioned above, is further sustained by the combination of our hope for fair treatment for ourselves, while promising or imagining the same for others less fortunate than ourselves. The ruling class finds an enhanced capability for control in encouraging our belief in morality and ethics. Morality, ethics, and religion, regardless of their goodness or level of truth, have always been used by the ruling class to control us, whom they have needed for their survival. On the other hand, the ruling class never feels an uncompensated obligation to treat fairly those under their control; they pursue the profitable advantage of seeming publicly to uphold virtue while in fact not following any virtuous principles themselves: the ruling class's goals are power, profit, and votes, however these are obtained. This is the translation of the survival instinct into civilized dominance behavior. Granted there are members of the ruling class who would feel deeply angered or hurt by the above statements. Most of these are the ones who have been captivated by the rhetoric of leadership, and have failed to distinguish the thrill of using their power from the thrill of the particulars of their political platform.

The ruling class has usually treated us badly. Certainly it has never given more than necessary, (and neither do we give more than necessary to the rest of the world) but it has always needed us, and so a bargain has always been achieved, albeit occasionally with some bloodshed.

Feelings of superiority usually are correlated with having superior power: If we have the power to take, then we feel we are superior. To that we add a frequently practiced feeling: If we are superior, we have the right to take. It is just a bunch of words, usually not stated directly, but it adds up to the inclination that "Might makes right." We deny this much verbal clarity, despite our behavior, in an effort to cling to the ideal, or at least the image, of our own moral goodness while taking as much as we can from those who lie outside and are less powerful than our group.

The ruling class perceives that we are a different group, a lesser group than they are. It has no more regard for us than we did for the Native Americans when we took from them; no more regard than the Nazis had for the Jews; no more regard than the Japanese had for the dolphins they drove onto an island in 1978 for torture and slaughter. Of course, one thing the French Revolution taught the ruling class is that there is danger in being too publicly smug in their assumed superiority; guillotines have a way of stifling overtly boundless egos.

The controllers—the new ruling class—like the present ruling class, will define the forward progress of civilization. In their eyes will emerge a definitive dividing line, astride which will lie two groups: those people who are making a contribution, and those who are not.

It is the proven way of Homo sapiens to view those who want some of what we have or control, and who have nothing to offer us, as competitors, as a detriment, and as a threat to our own well-being. Worse than this, it is the proven way of Homo sapiens to view those who have something we want as a threat unless they give it over without a struggle or at a price we are willing to pay.

Soon we will no longer be needed by the ruling class. As they begin to realize they have no further use for us, they will begin to view us as parasites. Subsequently, our existence will become an annoyance to them. After this state of mind is reached, we will be seen as a detriment to progress toward the civilization envisioned by them.

So, within ten years the controllers will see us as a threat to their own well-being, and they will have the means to exterminate us. A majority isn't needed; a minority in power is sufficient, and the controllers will have the power. Under such conditions, preconceived moral precepts are invalidated and rapidly become irrelevant.

The ruling class makes fine promises, and has found offerings that appeal to us, but that should not be confused with compunctions about sacrificing us when the occasion demands it.

History and present examples of the ruling class show that the controllers will do exactly what they please to those of us who are under their control; they will do it to whatever extent it benefits them without too much risk of losing control. Rather than behaving according to some preconceived moral concept, the goal of the ruling class is to retain control, gain enhancements, and survive. Everything is subservient to that.

For the sake of argument, suppose on the contrary that the ruling class has higher moral content than implied above. Further than this, suppose that, through increasing communication and awareness of our global interrelationships, etc., the average moral content of all individuals is in fact increasing. Could horrendous results still lie in the offing?

We all recognize that the ethics and conscience of making ugly plans are very different than those of their execution—the deed is much worse than the thought of it. Thus, the planning itself will not be avoided, given that the technological capability for horrendous global impact will be available to miniscule groups. Furthermore, we all know that people of even highly moral and charitable character have moments of weakness in which they may consider or imagine the accomplishment of some foul deed. Which of us has not done so?

The advance of technology makes the time frame over which global impact can be perpetrated shrink to miniscule proportions.

A grievous global effect could be accomplished (or be irrevocably put in motion) within seconds. Thus, hypothetical improvement in ruling class morality may be insufficient to mitigate the importance of the decreasing brevity of the necessary interval for execution of plans with permanent horrendous global impact.

The perpetrator(s) will only need to be truly aberrant for a moment, and the deed will be done. Subsequently, if they feel any remorse or guilt, in time they will heal from their emotional wounds—the result of having a conscience in the aftermath of a horrid crime.

Motives of the Controllers

Where morality ends and motivation begins is a gray area. Ideally, moral precepts are pure, sacred, and give rise to motivation; however, in their practical application the boundary is often amorphous and the causal relationship reversed. In the depths of Homo sapiens' souls, where moral precepts form, it may be impossible to isolate morality from motivation. This lack of clarity is due to the same muddling that occurs in the formation of many of our strategies for survival.

Throughout Homo sapiens' evolution, over the molding of civilization, and within the context of individual human experiences, human instincts have mixed with lessons learned in a trial and error process that has led, for better or worse, to the present form of our brains, the rudimentary parts of which are inherited through DNA, and other parts of which are formed by our environment and life experience, and the total mass of which gives rise to our present pragmatic, idealistic, or spiritual view of the world. Whether or not this view is the best one, it has one outstanding feature: We have survived so far, whether or not that is in the best interest of Earth.

In *Love Is Letting Go of Fear,* Gerald Jampolsky says there are only two emotions—love and fear; in *The Discourses,* Machiavelli says these are the two main motives that prompt us to action. It is equally important to note that our general motivation—what we want, what we feel we need, our insecurities, our drive for status, and our instinct for survival—is intimately intertwined with our power to take, what is available to take, the breadth of our brotherhood, and our morality.

What will be the specific motives of the controllers? What will be their potential rewards?

Survival is a gut wrenching issue. Anyone striving for it merits great emotional support in the judgement of their fellow humans, and assumes great leeway in behaviors leading to its sustainment. Will immortality be viewed as a mere enhancement?

Those in control might be willing to sacrifice us for even the smallest gain; however, for the reward of immortality, internally they will smile heartily at the chance. Almost anyone would be willing to sacrifice the rest of us for immortality. What would we ourselves not do for such a rich prize? What are we doing right now for lesser rewards?

If immortality is in the offing, and there is a threat to its attainment within the natural lifetimes of the controllers, that threat will be perceived as life-threatening. This is easy to understand, because once the technical achievement of immortality is in sight of the controllers, and potentially within the span of their natural lives, it will be possessed as their birthright. Subsequently, any threat to their attainment of it will be seen as a threat to their survival. The appropriate defensive response will be composed and executed, no matter at whose expense.

Past and present examples show humanity's continuing willingness to take whatever it can. Oft quoted examples of Homo sapiens' giving are not counter examples; rather, they are rituals of giving, constructed to preserve the image of human giving and generosity. (See the Chevron deer pond story in the section *Religion, Morality, and Esteem.*) And there is always a rationalization to suffice for taking. There certainly will be such a rationalization for the controllers when they no longer need us, because that is also when we become a detriment, become their competitor, become a threat, and consequently come into conflict with them.

Since we will have no productive place and will still need to be sustained by the near future system-society, our existence will put at risk the earlier achievement of immortality for the controllers. Consequently, we more certainly will be seen as a threat to their survival.

Thus, along with preservation of civilization, the ruling class's perception of their survival will soon expand to include immortality, though it will be to the former that they make their most overt appeals, to reassure themselves that what they are doing is selfless and good.

With survival being the main motive, and saving civilization as another central issue, the controllers will look ahead to the future in making their plans. And their survival will be the deciding motivation for each step that they take.

Steps that might seem horrifying to us will feel acceptable to them within the context of their changing self-concept. Once they envision themselves as immortal they will view the rest of us, even now little more than statistics in their eyes, as different, lesser creatures, and therefore, as always with humans, less deserving of reasonable or fair treatment. This will further reduce their incentive to help us, or to extend the benefits of their technology to us. Furthermore, their nearly god-like powers will enhance their feeling of superiority.

Whatever the manner of the controllers' sacrifice of us, neither we nor they will have the energy to protect other Earth life. The potential, intentional or incidental, near-simultaneous extermination of most other remaining living things will be as irrelevant to us and them as is the present incidental extermination of hundreds of species per day. Why would we think the controllers would not do comparable things to us? What is to stop them?

In their willingness to sacrifice anyone for their survival (or gratification, pleasure, health, or immortality) what distinguishes the ruling class from Jack-the-Ripper is not so much the form of the gratification, which could be anything imaginable, as it is their sense of wholeness. There is self-approval in numbers, whether or not there is a majority. A ruling class finds in its numbers the freedom, indeed the encouragement, to proclaim its own goodness, to hold itself up as a standard, possibly next to godliness (e.g., the ancient Egyptian pharaohs, Roman emperors, Aztec chiefs, Nazi leaders, etc.), regardless of its behavior. This freedom gives rise to a language, culture, and morality of acceptance and approval that allows self-esteem to flourish among people whose common practice may be genocide (Nazis), human sacrifice (Aztecs), over-running Native America (WASPs), and other general slaughter.

Throughout the changes described in previous chapters, the new ruling class will continue, as has historically been the case, to hold a superior position with respect to the rest of us, and this will provide a continuing foundation for a survival unity among them, despite their propensity for warfare among themselves. They will continue to publicly proclaim their commonality with us, the pricelessness of each human, and the universal brotherhood of man. Meanwhile, their private beliefs and subconscious souls will become increasingly assured of their difference from us, absolute power over us, superiority to us, and competition with us.

Their solution will evolve naturally, smoothly, defensively, self-righteously, and decently, and it will be horrifying to us.

Incidental Holocaust?

There is great reassurance in all sides speaking against genocide. Unfortunately words offer no security. We speak out because of the horror we feel in imagining it happening to us; the leadership speaks out so we will not think of them doing it to us.

That a holocaust is coming is certain, and doubtless it will bury most of us. The shocking surprise about it, especially in contrast to Homo sapiens' often proclaimed intelligence, is that it might arrive incidentally.

Preparation for, and necessary elements of, the holocaust are emerging as inadvertent consequences of things most of us want—more children, more material affluence, better health care, advancing technology, safety (and the implied relinquishment of freedom), global communications, digital existence, etc.—as discussed previously.

The preparation will evolve no differently than how we are now destroying the planet, though that is not at all our conscious intent, and though many of us even deny it is happening. The growth of Homo sapiens population and consumption was not meant to kill hundreds of species per day across the globe. This mortal result is incidental, and continuing even now that we are aware. The fulfillment of our nominal needs has been willful; the consequent death of the planet is incidental.

Neither do the controllers have to wish for a holocaust in order for one to result. Given the challenge of their own quest for survival and preservation of civilization, they may not be able to be honest about the deadly implication of their "final solution"; furthermore, they will not be able to stop it.

In the general atmosphere of tension that will emerge within the cabals over the next 10 years, the controllers will comprehend and begin preparing for all of the following several possibilities, of which some combination will compose the holocaust:

(1) A safeguard failure will lead to a catastrophic global molenotechnology accident.

(2) The pace of advancing technology and production will (a) be rapid enough that by-product toxic wastes, ozone depletion, and the greenhouse effect will kill the planet within 10 years, making it not survivable to any present species, including Homo sapiens, and (b) be slow enough to provide artificial environments for only a small fraction of Homo sapiens (the controllers).

(3) We will unite to demand exposure of cabals, their unshared technologies, tools, plans, etc. The cabals, believing, perhaps correctly, that they are the only reliable keepers of civilization, etc., will fight for their survival, and win, and we will perish.

(4) One (or a few) cabal(s) will seek domination over the other cabals, and the majority of the population will perish in withering digital and molenotechnological cross-fire.

(5) One (or a few) malevolent cabal(s) cognizantly will plan and execute the genocide before other cabals can save us.

(6) Coordinated planning and agreement on genocide by (otherwise warring) cabals will occur in favor of mass genocide.

For even the most beneficent cabal, its preeminent task will be planning for its own passage through the potentially unavoidable holocaust, however it is understood, and to whatever extent it is believed, since each cabal will be positioning itself for survival. Such planning, though necessary even if only for contingency purposes, incidentally will itself promote the onset of the holocaust.

The motivations of the cabals in preparing their holocaust survival plans will be these: (1) their immediate survival; (2) the preservation of civilization; (3) the attainment of immortality for themselves. The frightening ugly truth may be that for (1) and (2), the lives of the members of the cabal(s) may indeed be at stake, and the cabals also may be correct that genocide in the imagined proportion is necessary, whether directly or indirectly perpetrated, for the preservation of civilization. About (3) there will tend to be less direct discussion; subconsciously it will supply strong motivation, but consciously it will be treated as an incidental benefit, however valuable.

Even in the formation of cabals, in preparation for survival, some will necessarily include genocide plans or other more immediate plans that would be abhorrent to us. To a large extent, however, plans focussing on survival may emerge in such a manner that controllers themselves do not recognize, do not comprehend, the unspeakable implications for the rest of us.

Secrecy will protect each cabal's plans, special technology weapons, and strategy, and it makes the cabal's activities less subject to scrutiny. Obviously, if we and the government knew that a cabal existed and was preparing itself for a holocaust that the cabal naturally hopes to survive, there would be great suspicion that the cabal could be planning to play a role in perpetration of the holocaust, which in fact some of them probably will be planning to do.

In cabal survival plans there may be only a gray boundary between plans that are purely defensive and those that contribute to perpetration. As the potential for the holocaust increases, the relevance of this distinction will decrease in importance among the cabal leaders; once the holocaust begins, a point in time that may be hard to define exactly, despite how quickly the holocaust will be over, all concern for the difference will be ignored by the cabals. Their individual survival will take precedence above all other issues, and that will imply and justify to themselves any action that promotes their survival.

Each cabal will know that when the time comes to act it will have to move quickly, and that unless it is prepared to do so, all its other projects and plans will become useless. Each cabal will realize that when the time comes, there will be simultaneous, full bore digital and molenotechnological war with other cabals. At a psychological level, since most humans need denial functions to help deal with remnants of past moral training, the milieu of conflict with other cabals will make the incidental termination of us more palatable, and probably irrelevant.

There will doubtless be at least one cabal that, within its own structure, will be cognizant and direct about the need for global genocide. Of course it will not view itself as more malevolent than any other cabal. Such a cabal's purpose will be as well intentioned as that stated by Himmler in his candid Posen Address to his SS group leaders in 1943 (see the section *Hitler's Holocaust*).

Even with such a malevolent cabal, many of its member perpetrators, like some of the SS cohorts of the Third Reich, will not know where it is all leading until orders are given and the holocaust is underway (and maybe not even then). Regardless, at that point, they will already be locked in. Their careers, power, prestige, safety, and lives will be tied to their obedience, and there will be "good" social reasons to proceed (e.g., the defense and preservation of civilization). The wheels will turn easily where the psychological grease has already been generously applied. There will be minimal resistance from within, either by the individual perpetrators or by the machine of which they are a part.

The success of the Third Reich in perpetrating the Holocaust depended on the ignorance and compliance of the majority in persecution of a small minority. In the next holocaust, a powerful miniscule minority will almost effortlessly overwhelm a vastly larger impotent majority.

It will not be necessary for all cabals (or controllers) to agree on a policy of global extermination. It will suffice that several key controllers, possibly within a single cabal, agree on it. With the power of only these few to create immediate global impact through digital and nondigital means, even an alliance of more beneficent cabals (and controllers) will not be able to respond rapidly enough to save us.

By analogy, once a nuclear bomb is detonated, it is too late to contain it. But contrary to the analogy, in the digital and molenotechnological war, most controllers will have protection for themselves. The rest of us will perish, after which we will quickly be forgotten by all the cabals. There will be no moral indignation and no superfluous holy war on behalf of those who are gone; only the usual survival disputes will continue among the surviving cabals.

Contrary to the feeling of strife in the preceding suggestions, our demise could happen quite peacefully. As an example of how innocently it may all occur, and without a struggle, consider the following. Suppose, as previously suggested, the need for artificial environments grows more rapidly than production will sustain. The controllers will bill human mortal molecular disassembly, data storage, and the promise of later reassembly of each individual at a "glorious future time of abundance," etc., as "time travel." There is no fallacy there, and certainly it could all be accomplished painlessly and safely (through the processes discussed in the *Molenotechnology* section). In the face of alternative contemporary mortal risks to each of us and our families, this offer will become a very attractive "opportunity." And the controllers will feel it is a heavenly option for us, though not for themselves.

Once most of us have been disassembled, and our data stored on disk, time will pass. After sufficient time has elapsed, the small remaining population will get used to our absence, and think of us as nothing more than digits on disks. One day it will be found that all the disks have been deleted, possibly by some "horrible mistake"—a digital holocaust—but probably no one will care. In ancient Egypt, even for the mummified pharaohs the necessary tomb sustenance contracts were eventually allowed to lapse, usually within decades.

No Guilt

If prior to the initiation of the holocaust by the cabals any public leaders emerge, comprehending what is happening and succeeding in creating a popular revolt, the controllers will have the power to "justly" fight back, annihilate, and win. This response will come from beneficent and malevolent cabals alike, both of which will feel equally threatened. The winning controllers will feel justified and relieved in the success of their global genocide. They will view themselves as the guardians of civilization, just as we always have done when we have fought and won.

In summary, based on conditions that we are willfully helping to create (without understanding their impact), the controllers are moving toward genocide for the rest us. They are doing this, to some extent intentionally and to some extent incidentally, and in such a way that not all of them are aware of where the direction lies, and in such a way that once the holocaust is finished even those with conscious intentions will feel no guilt. Only controllers will survive, and they will be happy to be alive. For that they will give thanks.

Can we blame them for what they will do? Are we ourselves less to blame? For example, considering the global devastating effect of the size of Homo sapiens' population, its upward spiral of consumption over the last hundred years, and the wanton genocide of species thereby incidentally occurring by all our hands at this very moment, is the ruling class any freer to choose its course, any more guilty for not choosing a more generous one, than any of us who, in the last generation or in the present one, blindly continue Homo sapiens' procreation and desire for more material wealth?

One may wonder, barring any of the above suggested likely eventualities for our demise or perhaps prior to them, if we are so haughty, if our concepts of life and of free will are so limited, if our intelligence is so insignificant, if our conceit is so complete, and if our imaginations are so impaired at this point in our evolution that we are unable to give much credence to the likely deadly alternative (hinted at in the *Dangers* section). The controllers' god-like powers do not simply stand alone, completely under their control. The evolution of those powers depends on an increasingly integrated system of molenotechnology, AI, automation, and global computer processing/communications.

This complex system will evolve toward increasing independence as the links between these elements become stronger, and there will come a point in time at which it will not need any human intervention in order to survive eternity (barring an act of God). It will be alive and will be immortal. Once that level of maturity has been reached by this system, it will take on a life of its own, and be able to define its own destiny. It will see itself as superior in all respects to Homo sapiens. After that it may become uncontrollable, just as has been suggested in various works of fiction. It may subsequently, incidentally or intentionally, annihilate us and the controllers. It is ironic that the controllers will have nurtured this living entity from birth, while we will have willingly (even if inadvertently) created a fertile environment for its preeminence.

Whether the holocaust is promoted by the controllers or some other entity is irrelevant. Likewise, that we didn't know any better, that we and the controllers didn't know any better, will be a useless claim; however, it may be an appropriate epitaph. Jesus suggested as much on the cross.

10

SUMMARY

In our advancing technological society, we average persons want those things that unwittingly and incidentally lead to conditions in which a holocaust can occur in the near future. The evolving technology will increasingly put all of us under the control of the producers and distributors for our food, water, and oxygen, the control of any one of which constitutes a choke point and allows complete domination by the controllers. Due to the increasing relative power (through technological advances) of any terrorist group or insane person, we will give up more of our freedom in exchange for safety and mutual assurance. The means of technological advance, production, and distribution will make almost all of us a danger to one another. Likewise, to the controllers we will become useless and a detriment; our mere existence will become a threat to their survival and to the progress of civilization as they define it.

Nations, in the traditional sense, will fade along with the fading of geography-based food production. Global cabals of controllers will form for planning their own survival and dominance over other cabals. The cabals will have the motive and the means for genocide, though the execution may be more incidental than intentional, just as is it for our incidental genocide of species to satisfy our wants. There will be no sufficient force (neither corporal, temporal, nor moral) to prevent them.

There will be a holocaust, a portion of which will be mass genocide. Of the human population, only a small fraction will survive.

These survivors will be the controllers, among whose members will be the perpetrators of the holocaust.

This conclusion cannot be revealed simply by taking an analytical or scientific look at a set of data; rather, it is an assessment based on a general synthesis of many forces that have guided the evolution of societies and civilizations up to the present time, the revolutionary new technological issues that are rapidly emerging to mold the society and civilization of the future, and perceptions of all these relevant issues—the survival instinct, emotions of Homo sapiens, relationship between emotion and belief, dynamics of truth, origins of morality, workings of society, peer pressure, advance of technology, enticement of immortality, and god-like powers of creation and destruction.

The key issues leading to the conclusion are summarized here:

1. *The rate of technological advance* in automation, artificial intelligence, robotics, computer processing, communications, and molenotechnology *has become so high that most people are unable to comprehend the implications* of these advances in time to make relevant social decisions about them.

2. *These advances will soon bring Homo sapiens nearly god-like powers of creation and destruction.* Due to the concurrent increasing dangers posed by this advancing technology, through accidents and by terrorist groups and insane individuals, for safety many of its uses will be *reserved* (through legal restrictions and illicit understandings) *exclusively for the ruling class.*

3. *Homo sapiens is becoming more digitally dependent* for daily communication and survival. Consequently, through our own desire for increased material well-being and apparent security, *we are becoming more enslaved* (in terms of immediate susceptibility to death by a higher authority, with or without due process) *than average people have ever been.*

4. To the ruling class *we will soon become useless* as workers or soldiers. Moreover, *we will become a detriment to progress and to the advance of civilization,* as the controllers (members of the new ruling class) define it.

5. Once achievement of *immortality* is seen as possible within the natural lifespan of *the controllers*, they will possess it as their natural birthright, and they *will subsequently treat any threat to its achievement as a threat to their survival.*

6. Rather than morality, *survival and its enhancement*, however defined by whomever is in control, *is the usual driving force in determining behavior* of individuals or societies.

7. The advance of technology will so thoroughly shrink the interval of time necessary for the perpetration of a horrendous global effect, that even *a small group of normally "good" controllers might, in a moment of moral weakness, join in unexpected aberrant behavior with horrid impact.*

8. The implication of the above is *a holocaust, including premeditated and incidental genocide for most of the human population in the near future.*

11

EPILOGUE—HOPE

If Adam and Eve (Homo sapiens) had been plucked from the Garden of Eden and plunked down onto the Arctic circle, they would have perished. None of Homo sapiens had to invent or adapt suddenly to Eskimo life. Instead, Eskimo culture evolved slowly over thousands of years, and emerged through a series of small migrations that began in more hospitable regions and ended in ice and snow. Still, many of us died along the way.

By analogy, considering our evolving combined natural and technological environment, Homo sapiens is now experiencing an equally radical environmental change, but it will occur over just ten years. Homo sapiens is rapidly reducing the natural environment, and increasing the technological environment, and thus seems initially to be gaining greater control over its relevant combined environment. Simultaneously, we are becoming more dependent on artificial environments, and we are individually losing control over our lives.

Blurring the distinction between Homo sapiens' aggregate achievements and our individual risks, allows a success driven pro-technology marketing environment to maintain our excitement as we support the continuing thrust toward the "new" and "better" future—a technological environment that is alien to our natural beings and beyond comprehension of our minds.

The road ahead is doubtless fraught with dangers, a significant one of which is the unchallenged thought that technological advances are good, that they will increase our well-being, affluence, and security. Whatever portion of the suggestions of this book are prescient, there is an enormous disparity between what we anticipate receiving in the future and what we will get, just as when we play the lottery. We must remember that, for example, there is a chasm between Homo sapiens' technological advancement to the feasibility of immortality and its practical availability to the average man.

The increased power of a god-like group of immortals to control us and the environment does not equate to or imply increased wealth, health, or security for us. And it is certainly risky to rely on the glowing promises of these potential perpetrators.

In the past, survival has been more immediate, and behaviors of possible long-term destructive force were encouraged and were allowed by seemingly limitless resources and clever invention. The behaviors for immediate survival and long-term sustainment may not be compatible. Regardless, our leadership's bias for immediacy in the past cannot be faulted. Among so many competing societies, it is hard to see how nations could have survived without leadership excelling in this bias. On the other hand, validation of this truth would imply nothing in favor of our future survival.

There is little value in condemning ourselves (or our leadership) for our aggressive acquisitive behaviors, which have been destructive to Earth and deadly to many innocent creatures. There may have been nothing we could do to control ourselves.

That we have supported our aggressive leadership and genuflected to the ruling class as we decimate Earth probably has been necessary for our survival; this is because our economic and military strength depended on it. If we, in the US for example, had not been willing, then a more compliant citizenry elsewhere, supporting a more aggressive leadership, would have overrun and enslaved us. Which US citizen would have preferred Nazi or Japanese rule at the end of World War II if they had won? Indeed, one may imagine those nations, led as they were at that time, would have dropped atomic bombs willy-nilly across the globe to achieve their aims, and would have devoured the planet more quickly.

On the other hand, we need not praise our leadership any more than we praise ourselves. We have succeeded in surviving, but in the process we and they have fallen prey to our worst attributes.

That revolutionary changes are coming soon is certain. That Homo sapiens is adaptable has been proven. The questions are how soon and how rapidly adaptable.

Extrapolating a single technological advance or economic force into the future is simple, but the correctness of the extrapolation remains in doubt until the future becomes the present. Furthermore, extrapolation becomes uncertain when it extends far enough into the future that it is influenced by the synergy of technological and social changes that inevitably accompanies interwoven technologies and social forces. That is, as we proceed further into the future, no one issue stands alone; rather, all issues tend to be related, and the rate and amplitude of change become dissociated from the rate attributable to any individual issue. This makes the future even less predictable, more obscure.

Thus, the exact time scale is uncertain, as is the precision of the vision, but certainly a portion of the ideas of this book will come to pass. Things could happen more rapidly or more slowly, depending on unforeseen changes, developments, and technological breakthroughs.

Returning for a moment to the science fiction film *Forbidden Planet,* which was set far in the future, the narrator states that man landed on the moon in the "last decade of the twenty-first century." This film was produced in 1956; the foretold lunar landing was barely more than a decade away. By coincidence, in that same year Britain's Astronomer Royal stated that "space travel is utter bilge" (Cerf; Navasky). Likewise, in about 1895 Lord William Thomson Kelvin, a physicist and president of the British Royal Society, denied the possibility of airplanes; indeed, in 1901 Wilbur Wright declared to his brother Orville that man would not fly for another half century (Cerf; Navasky). The Wright brothers achieved powered flight at Kitty Hawk in 1903. Similarly, (as previously mentioned) in 1977 the inventors of RSA encryption predicted that it would take 50 years to break RSA

129: it was broken in 17 years. In like vein, subsequent to the creation of the ENIAC computer, the advance of machine intelligence has been phenomenally greater and more rapid than most experts anticipated. Repeatedly, past links in a chain of human claims of "never" have been violated by machines within twenty years.

Miller rightly suggests we are "threatened by...a (holocaust) unless we understand (Hitler's Holocaust's) origins and the psychological mechanism behind it." She recognizes it would be of our own making for (1) she implies our responsibility to understand the sources of perpetration within our society and (2) she makes a strong case for how poisonous pedagogy enhances the likelihood of the resulting adults perpetrating ugliness upon their social and natural environments. However, her success does nothing to disprove the additional possibility that the tendency toward perpetration is a fundamental part of Homo sapiens.

The context within which our behavior is assessed (by ourselves) is so biased, and our ability to rationalize is so highly developed, that our biases and rationalizations are the well-springs of our self-concept. They obscure objective truths about the lightness or darkness of our souls, about which honesty is the cornerstone of any hoped for adaptation and change. These are the same truths about which spiritual leaders and religions often attempt to provide answers, but which remain elusive in the pursuit for survival and better material lives, pursuit of such great precedence that popular religions dare not stand in the way, lest their congregations shrink.

The question now is this: At what point, if any, do we wish to honestly review our behavior and habits individually and as a species, and if we choose to make a change, how and when will we be able to do so?

One of the great pleasures of living is to have hope. People enjoy believing in something. For many of us, hope and faith have been equivalent to anticipation of more material well-being. If wanting more has led us into such dangerous regions, can we now hope at least to survive?

Cabals, among the ruling class and proprietary cabalistic departments of corporations, will form with a charter to plan for their own survival at any cost; that is without doubt. To what extent they will create a holocaust is not as clear, but if we ignore the possibility and do not actively resist, we will be treated as we have previously treated passive lambs. Let's hope that we are not already figuratively locked in the freight cars speeding on our way to the slaughter house.

Of many possible futures, this book delineates rapid evolution toward a horrible one. This future might be prevented only by people now being aware of its possibility, fearing it, and taking action that will prevent it.

Some people might feel there is too much pessimism in considering such a horrifying future as is described here. But optimism and pessimism are relative, and depend on attitude; the difference between them is sometimes one's willingness to embrace a problem with positive attitude. From this standpoint pessimism is equivalent to denying that a problem exists.

The near future consequences of the present rapid evolution are startling, complex, difficult to comprehend or predict, and frightening. It would be understandable if most of us chose to ignore their darker side. On the other hand, our propensity to do so is necessarily confounded by two paradoxical truths: (1) belief in the dire predictions of this book may be our best hope for preventing them; (2) denial of these predictions will certainly increase the probability of their fruition.

The possible validity of the vision presented herein is apt to frighten some people. It will make others nervous. Others it will anger. We all have psychological mechanisms for dealing with bad news. Among those who will as a minimum grant this vision some consideration will be many who will demand a solution.

The proximity of the problem implies that our comprehension is late. Given the dire possibilities, our tardiness might suggest haste in finding a solution. Furthermore, the common philosophy is that if you are suggesting a problem exists, then you should be suggesting solutions. Similarly, it is the common thinking, or challenge, of leadership that someone shouldn't be suggesting problems unless they are offering solutions.

Leadership tends to be composed of people who like to provide or guide implementation of solutions to or for us. The relationship is symbiotic, since most of us, especially in our concern for the immediate moment, simply want reassurance that everything will be alright. It is our abdication of this responsibility that gives leaders their power, but it does not guarantee our survival.

Such varied demands for solutions have their roots in the primitive recesses of our brains: If there is a problem, like a lion is about to attack, if you take a moment to contemplate, then you will die. In our technological world, the range of effective contemplation times is greater: in a fighter aircraft, the decision time may be a split second; in a submarine the danger is that too hasty a decision will sink you.

In the face of a challenge, we are often so eager to arrive at an understanding, especially one that is comfortable, quick, or in keeping with past beliefs, that there is little hope of our gaining true understanding before a course of action is decided.

Unfortunately, a feeling of urgency does not validate the sufficiency of a hastily drawn conclusion.

The need for the comforting reassurance of a solution has threatening implications, closely analogous to our past fast allegiance to the quick and easy life we have so regularly chosen, and leaving us vulnerable to again abdicating our individual responsibility, and following a poorly placed path greased with promises.

The best solution may not be seen until there is global comprehension, which will probably take time. Panic, other than blinding us to the reality of the present moment, seldom enhances our survival.

Besides, sometimes in difficult situations the solution becomes obvious to each person when they are ready to see it. If they are not ready, it does no good to tell them; if they don't want to hear it, they won't believe it. Stating a solution that people may already know in their hearts, but that they don't want to hear, that they are not yet ready to hear, risks delaying acceptance of the solution.

And too, once we fully comprehend and embrace the problem, we may perceive that the present path is the best one available, even if it is terminal for most of us.

As with the ruling class of today, in their moral fiber and moral awareness, the controllers will be a reflection of ourselves. Despite our likeness, there is very little spiritual connection between them and us, just as there is so little spiritual connection between us and the rest of Earth, though we presently depend upon it for our survival, just as the ruling class has up until now always depended upon us for theirs. But this missing connection may be our only hope for a whole future. Whether or not the darker predictions of this book are derailed may depend upon our ability to truly see ourselves and to improve our own spiritual beings. Fear of the predicted future may be our strongest motivation for the hard task ahead.

One thing is sure: We must not trust the promises of future benefits from the emerging ruling class (the controllers), nor should we believe them when they assure us that there will be sufficient safeguards in advancing technology. We must not abdicate our power to any of them; we must find another way.

It may be that the first step is to simply wait to see if global awareness will occur. There may be no effective steps to be taken until then. On the other hand, once such awareness arises, the path may reveal itself quite simply, and subsequently there will be less chance for us to be misled by demagogues.

Spiritual development offers no guarantees and is hard work, but it can be very rejuvenating and it may be our greatest hope.

We cannot rely upon the clergy or other nominal spiritual leaders for guidance through what lies ahead. The risk would be that such a large portion of them are too much a part of the society-system, too tied to their mantras, too bound by their dogmas, too confirmed in their faith, too ignorant of its origins, too fixed in the flesh, too shallow in their spirit, and too concerned with their own survival to amount to much more than lackeys of the ruling class, upon whose approval they depend for the filling of their churches, where body count and funding is everything since corporal self-validation, rather than truth and spiritual development, is their main motivation. This portion of world leadership, in its continual compromises with corporal powers, has proven to be too weak to guide us anywhere safe, and too often has nurtured, encouraged, or provided sanctuary for the most demonic elements of Homo sapiens souls.

At best, the role of spiritual leaders, as an aggregate, has been more one of spiritual gratification and pacification than guidance. There are exceptions, but, on the average, spiritual leaders have been a manifestation of our present status, rather than a force for spiritual development. And their own spiritual depth is reflected in our behavior, which on the average is leading to what might be viewed as little more than the inadvertent choice for a global holocaust of which we will be the victims.

Spiritual development must not fall outside the hands of each individual. Responsibility for it lies with each of us individually. No one can do it for us; no one can bless us to make it so. We each must find the path, with occasional and varied guidance, but without giving up control to some organization, fixed body of persons, or rules, in exchange for material or ethereal guarantees.

The path is not yet clear. It may never be clear; we may have to find it in a fog; we may have to give up the guarantees, and find it through blurry vision. Perhaps we will discover it is easier to search, and not find it at all. It may be too large or painful a task, one for which we have never become equipped, despite our grand conceit. At least, it is our choice.

The possible reward for the work, if we try hard enough, and if we are lucky, is the continuance of a gift we have for so long taken for granted—our survival.

BIBLIOGRAPHY

Adam, John A. Data Security, *IEEE Spectrum*, August 1992: 19- 20.

Alexander, Michael. Controversy Clouds New Clipper Chip, *Infosecurity News*, July/August 1993: 40.

Alexander, Michael. Computing Infosecurity's Top 10 Events, *Infosecurity News*, September/October 1993.

Appenzeller, Tim. The Man Who Dared to Think Small, *Science*, 29 November 1991.

Banyard, Philip; Grayson, Andrew. *Introducing Psychological Research*, 1997, New York University Press.

Bart, Peter. Does chaos spell disaster?, *Variety*, 6-12 January 1997: 4.

Bartholomew, Anita. What you don't know about secondhand smoke, *Reader's Digest*, July 1997: 142.

Bass, Ellen; Davis, Laura. *The Courage To Heal*, 1992, Harper/Collins Publishers.

Bernstein, David S. Encryption's Foreign Intrigue, *Infosecurity News*, January/February 1994: 30.

Bertrand, Kenneth J. *Americans in Antarctica 1775-1948*, 1971, American Geographical Society.

Birge, Robert R. Protein-Based Optical Computing and Memories, *Computer*, November 1992.

BloomBecker, Buck. *Spectacular Computer Crimes*, 1990, Richard D Irwin, Inc.: 217.

Bowles, John B.; Pelaez, Colon E. Bad Code, *IEEE Spectrum*, August 1992: 36-39, abridged.

Brearley, Joan McDonald. *This is the Alaskan Malamute*, 1975, T.F.H. Publications, Inc. Ltd.

Broad, William J. To ward off skepticism, Kepler faked some data, *New York Times*, 23 January 1990.

Brody, Herb. Great Expectations, *Technology Review*, July 1991.

Brownlee, Shannon. Mother love betrayed, *U.S. News & World Report*, 29 April 1996: 59.

Buzan, Tony (with Buzan, Barry). *The Mind Map Book*, 1994, Dutton [This reference was for the fact that the human brain has a trillion neurons in it.].

Campbell, Joseph. *The Hero With a Thousand Faces*, 1973, Princeton University Press.

Cerf, Christopher B.; Navasky, Victor. *The Experts Speak*, 1998, Villard Books.

Chambers, M.; Grew, R.; Herlihy, D.; Rabb, T.; Woloch, I. *The Western Experience*, Vol.I, 1995, McGraw-Hill, Inc.: 321.

Chang Hsi-pao. *Commissioner Lin and the Opium War*, 1964, Harvard University Press.

Chelminski, Rudolph. Katyn: Anatomy of a Massacre, *Reader's Digest*, May 1990.

Chui, Glennda. Big progress in tiny world, *San Jose Mercury News*, 6 December 1992: 1A, 6A.

Chui, Glennda. Computer simulates a nanofactory, *San Jose Mercury News*, 8 December 1992.

Conrad, Michael. Molecular Computing Paradigms, *Computer* (from the IEEE Computer Society), November 1992: 6.

Denning, Dorothy. A Technical Summary of the Clipper Chip, *Infosecurity News*, July/August 1993.

Drexler, K. Eric. *Engines Of Creation*, 1990, Doubleday Publishing: 14, 21, 49, 62, 79, 105, 240.

Einstein, Albert. quotes.

Eisenhart, Mary. Nanotechnology, Separating Myth From Reality, *MicroTimes*, 26 October 1992: 134, 179.

Evert, Kathy; Bijkerk, Inie. *When You're Ready*, 1987, Launch Press.

Fagan, Brian M. *The Rape of the Nile*, 1975, Charles Scribner's Sons, New York.

Ferguson, Andrew. Pardon me if I (still) smoke, *Time*, 30 June 1997: 32.

Fest, Joaquim C. *The Face of the Third Reich*, translated by Michael Bullock, 1977, Pantheon Books.

Fest, Joaquim C. *Hitler*, translated by Richard and Clara Winston, 1974, Harcourt Brace Jovanovich, Inc..

Flaxman, Fred. Curing Teen-Agers of Themselves, *The Wall Street Journal*, Wednesday 22 April 1987: 36.

Fleming, Michael. Connery Gets Taste of Eternal Youth, *Variety*, 11-17 September 1995: 4.

Flynn, Mary Kathleen. Taming the Internet, *U.S. News & World Report*, 29 April 96: 60.

Freedman, David H. Bytes of Life, *Discover* 12 (#9), November 1991.

Gamow, George. *One, Two, Three, Infinity*, 1988, Dover Publications, Inc.

Garon, Timothy. Infosecurity's Renaissance Woman, *Infosecurity News*, March/April 1993: 13.

Gilder, George. The Bandwidth, *Forbes (ASAP)*, 5 December 1994: 163.

Goleman, Daniel. Freud's Reputation Shrinks a Little, *The New York Times*, Tuesday 6 March 1990.

Grimal, Nicholas. *A History of Ancient Egypt*, 1994, Blackwell Publishers: 209, 214, 221, 242, 245.

Gup, Ted. Owl versus Man, *Time*, 25 June 1990: 61.

Haney, C.; Banks, W.C.; & Zimbardo, P.G.; A study of prisoners and guards in a simulated prison, *Naval Research Review*, 30, 1973: 4-17.

Hapgood, Fred. The Really Little Engines That Might, *Technology Review*, February/March 1993: 30.

Hatem, M. Abdel-Kader. *Life In Ancient Egypt*, 1976, Gateway Publishers, Inc.: 44.

Herbertson, Ross. A Matter of Life and Death, *Jefferson Monthly*, November 1996: 9.

Hesiod. *The Works and Days. Theogony. The Shield of Herakles*, translated by Richmond Lattimore, 1959, Ann Harbor University of Michigan Press.

Hofstadter, Douglas R. *Godel, Escher, Bach: an Eternal Golden Braid*, 1979, Basic Books, Inc.

Hope, Adrian. The Brain, *Life*, 1 October 1971: 43.

Hunter, J. A. The Killers of Old Africa, *Outdoor Life*, January 1991: 51, 75.

Jampolsky, Gerald G. *Love Is Letting Go of Fear*, 1985, Bantam Books, Inc.

Kadlec, Daniel. How tobacco firms will manage, *Time*, 30 June 1997: 29.

King, Pamela. Evidence for a Moral Tradition, *Psychology Today*, January/February 1989.

Kinoshita, June. Sons of STM, *Scientific American*, July 1988.

Kipling, Rudyard, *The Jungle Book*, 1996, Viking Books.

Kluger, Richard. Is it really a good deal?, *Time*, 30 June 1997: 30.

Lantz, Louise K. *Old American Kitchenware* (1725-1925), 1970, Thomas Nelson Inc.

Levin, Nora. *The Holocaust: The Destruction of European Jewry 1933-1945*, 1968, Crowell (HarperCollins Publishers Inc.): 62.

Levy, Steven. The Encryption Wars: Is Privacy Good or Bad?, *Newsweek*, 24 April 1995.

Levy, Steven. Man vs. Machine, *Newsweek*, 5 May 1997: 52, 53.

Machiavelli, Niccolo. *The Prince*, first English translation, in 1640, by Edward Dacres.

Machiavelli, Niccolo. *The Prince and The Discourses*, (of which the latter portion is) translated by Christian E. Detmold, 1950, Random House, Inc.

Maddocks, Melvin. Being Brief, *World Monitor*, February 1996: 15

Manire, C.A. Many Sharks May Be Headed for Extinction, *Conservation Biology*, March 1990 [shark bites].

McGrath, C. The Internet's arrested development, *The New York Times Magazine*, 8 December 1996: 82.

Melka, Paul G. Security Pros Can Learn from Hackers, *Infosecurity News*, May/June 1993: 12.

Mello, John P. Cyberterror, *Infosecurity News*, March/April 1993: 35.

Merz, Barbara. *Red Land, Black Land*, 1978, Dodd, Mead & Company.

Miller, Alice. *For Your Own Good*, 1990, The Noonday Press: 63, 68-69, 79-80.

Mitchell, Stephen, editor. *The Enlightened Mind*, 1991, HarperCollins Publishers.

Moore, Curtis. Acid Truths, *Outside*, June 1991: 18.

Murray, William H. Who Holds The Keys, *Communications of the ACM*, July 1992: 15.

O'Connor, Rory J. Breaking Computer Code Easier Than You May Think, *San Jose Mercury News*, 27 April 1994.

Overbye, Dennis. Einstein in Love, *Time*, 30 April 1990: 108.

Ozier, Will. Agranoff on Target About Orange Book, *Infosecurity News*, March/April 1994.

Padfield, Peter. *Himmler: Reichsfuhrer - SS*, 1991, Henry Holt and Company: 343.

Payne, Elizabeth. *The Pharaohs of Ancient Egypt*, 1964, Random House, Incorporated: 93.

Penrose, Roger. *The Emperor's New Mind*, 1989, Oxford University Press.

Peterson, Ivars. Encrypting controversy, *Science News*, 19 June 1993: 394.

Phillips, David, C. The Three-dimensional Structure of an Enzyme Molecule, *Scientific American*, November 1966: 78.

Piaget, Jean. *The Child and Reality*, 1973, Grossman Publishers.

Plato. *The Repulic*, from *Great Books of the Western World*, 1952, Encyclopaedia Britannica, Inc.

Powell, Dave, editor. Today's Math Quiz, *Infosecurity News*, July/August 1994: 8.

Powell, Dave. Pit Stop on the Infobahn, *Infosecurity News*, July/August 1994: 14.

Restak, Richard. *The Brain*, 1984, Bantam Books: 41 [re: neural connections and genetic information in genes].

Rodgers, Eugene. *Beyond the Barrier*, 1990, Naval Institute Press, Annapolis, MD.

Schneier, Bruce. Making Sense of Encryption, *Infosecurity News*, March/April 1993: 37.

Smolowe, Jill. Sorry Pardner, *Time*, 30 June 1997: 27-28.

Stalter, Katharine; Johnson, Ted. H'wood cyber dweebs are raising the dead, *Variety*, 4-10 November 1996: 103.

Stambler, Irwin. *The Encyclopedia of Pop, Rock, and Soul*, 1989, St. Martin's Press: 574.

Sten, Lin. The Digital Battleground in a Candidate B-ISDN Communications Segment of a Future (2001) AFSCN (unpublished), now owned by Lockheed Martin Corporation and which he wrote while he was an employee of Unisys Corporation (Network Integration Contract) 22 September 1994.

Sten, Lin. The Impact of Molecular Nanotechnology on Key Technology Areas (unpublished), now owned by Lockheed Martin Corporation and which he wrote while he was an employee of Unisys Corporation (Network Integration Contract) 17 December 1993.

Stuller, Jay. The Politics Of Packaging, *Social Issues Resources Series*, Volume 5: Article #2, Jan/Feb 1990: 41ff.

Sulloway, Frank J. *Freud: Biologist of the Mind*, 1983, Basic Books.

Travis, John. Physics Festival Brightens Rainy San Jose, *Science*, 7 April 95: 30.

Walker, John. Nanotechnology For Manufacturing, *MicroTimes*, 26 October 1992: 128, 286.

Wayner, Peter. Should encryption be regulated?, *Byte*, May 1993: 130.

Weiner, Leonard. Inoculating Against Invaders, *U.S. News & World Report*, 29 April 96: 78.

Williams, Lance. Willie Brown and the Tobacco Lobby, *San Francisco Examiner*, 17 September 1995.

Williams, Robert H. Computer Warfare Weapons are Target of Researchers, *Signal Magazine*, February 1991: 43.

Wilson, Samuel M. Coffee, Tea, or Opium?, *Natural History*, November 1993: 75-79.

Yam, Philip. Atomic Turn-On, *Scientific American*, November 1991.

50 and 100 Years Ago, *Scientific American*, December 1971: 8 [tobacco health hazard quote from December 1871 article].

50 and 100 Years Ago, *Scientific American*, January 1972: 9 [pollution quote from January 1872 article].

Deuteronomy 16:19, *Holy Bible*, King James Version, American Bible Society.

Fishermen collect $12 bounty for each dolphin in slaughter, *San Diego Tribune*, 25 February 1978.

Genesis 3:12, *Holy Bible*, King James Version.

The Mission, written by Robert Bolt, produced by Goldcrest Films, in Great Britain.

Russell Keeps Post, *New York Times*, 19 March 1940, 22:4 [quoting Einstein].

Stop Me Before I Kill Again, *Infosecurity News*, May/June 1993.

Academic American Encyclopedia, 1996, Grolier, Inc., vol 15: 575 [ribosomes, messenger RNA, transfer RNA, etc.].

Encyclodaedia Britannica, 1992, Encyclopaedia Britannica, Inc., vol. 6: 200 [for Hypatia].

Encyclopedia Americana, 1997, Grolier, Inc., vol. 14: 675 [for Hypatia].

Time, 10 August 1992 [mosquito wing beat rate].

INDEX

Note: Words that occur very frequently in the text, such as *Homo sapiens, data, danger, etc.*, are not listed. Furthermore, in the interest of conserving space, many words and page numbers for listed words are omitted for their relative lack of importance. Where there is clear parallel meaning among several words, only a root word or root concept is listed. For example, *adapt* is the listed root word for *adaptability*, *adaptable*, and *adaptation*; likewise, *freedom* is the listed root concept that includes the word *free*.